Understanding 3D Printing

Al Williams

1

ISBN: 150061727X
ISBN-13: 978-1500617271

DEDICATION

Always, for Pat..

CONTENTS

Note: For full color figures visit http://hotsolder.com/u3dpfigs

INTRODUCTION

I'm not sure what the watershed moment for 3D printing was: the president mentioning 3D printing during the State of the Union address, or the Big Bang Theory episode a few weeks earlier where two of the guys decided to buy a 3D printer to make custom superheroes. Either way, the message is clear. Everyone's talking about 3D printing.

Since you are reading this book, you're curious about 3D printing. This book will help you answer your questions and help you get started with 3D printing in your home or office. Of course, getting started will mean different things to different people. You will probably want to buy or build your own printer and that can cost anywhere from a few hundred dollars to the price of a luxury car. However, there are companies that will do prints for you, on printers you are not likely to want to buy yourself. Or you might find a local group who has a printer and you can use it for a membership fee.

Once upon a time—say around 1978—if you had a computer in your home you were one of a very select few. People would be curious about your computer and what you were doing with it. Today, no one cares if you have your own computer. They'd probably be more surprised if you didn't have one. It is going to be awhile before 3D printers are that popular, if they ever are. However, it won't be long before they are as common as, say, drill presses. You might not have a drill press in your garage, but you probably know someone who does. In any event, if someone told you they had a drill press, you wouldn't really think much of it. The 3D printer is well on its way to being that common.

This book will cover the mechanics and the electronics of printing. If you want to print things you download from the Internet, that's all you'll need to know. You'll eventually want to create your own objects and you'll need to use software on your PC to create models using Computer Aided Drafting (CAD) software or 3D modeling software like Blender. A complete treatment of CAD software is beyond the scope of this book, but you'll get a taste of some simple modeling techniques at the end of the book.

Do you need to know CAD? No, but if you do, it won't hurt. You should be moderately comfortable with a computer, though. Some CAD and modeling software will be familiar to you if you've ever used a paint program. Others will be easy if you are familiar with programming.

If you decide to build your own printer, you should be reasonably handy with simple tools. You don't need to be able to machine high-pressure stainless steel vessels. But you probably ought to be comfortable building some bookshelves from the local hardware store. If you don't have some simple hand tools, you are going to need some, but nothing too elaborate. You can even buy pre-assembled printers, but you'll still need a few simple tools.

You don't absolutely have to buy a printer or a kit. There are many printers with complete plans freely available. However, unless you are very technically astute, you would probably be better off buying a kit first, even if it is a small one and you'd like to build a larger one in the future. If you don't know what an ACME nut is or you don't have access to taps, dies, and a drill press, you probably want to start with a kit or an assembled unit. There are a few parts of a printer that you might elect to buy readymade, even if you are adept with tools. For example, the hot end that melts the plastic can be difficult to fabricate and many people will buy one already built, or at least a kit with the difficult parts assembled.

The same is true of electronics. Most people will buy a ready-built board although if you are comfortable soldering you could build your own. Some people who are accustomed to soldering prefer to build their own cables, for example, while others elect to buy ready-made cables and skip soldering.

In any case, you'll need a computer. Most software you need will run on Windows or Linux. Much of it will also run on the Mac, but this book will focus on Windows and Linux even though Mac users will probably be able to fill in the blanks.

People of all ages, educational backgrounds, and walks of life are using 3D printers. This book will help you join them.

1 WHAT IS 3D PRINTING?

Baby boomers grew up watching movies that showed the future with robots—Robbie, Rosie, and a host of other humanoid servants. Other than a few novelty robots, we don't really have anything like that today. But we do have robots—they just don't look like us. Little disk-shaped robots clean the floors. Robotic welders make our automobiles. There are even robotic lawn mowers.

One class of robots convert computer-generated designs and create solid objects from them. In reality, there are at least two types of robots that do this: Computer Numerical Control (CNC) machines are subtractive machines. They start with a solid block of material and remove the parts that don't belong. CNC mills and lathes have been common in machine shops for years.

Although some people have CNC machines at home, there are several reasons they are not more common. First, there are limits to what you can build with CNC machines. For example, you can't machine a hollow ball by removing material with a mill. There is no way for the tool to get to inside material. The other issue is how hard it is to remove certain materials. On YouTube, for example, you can find some CNC tools built out of children's construction bricks. They can only mill shapes into soft floral foam. It is a little harder to mill soft wood and plastic—it requires a stronger machine. If you want to mill, say, mild steel you will need a much stronger machine. Stronger machines are harder to build and also cost more money to build.

The alternative to subtractive machining is additive machining. These machines build up objects by placing material where there isn't any. There are actually several different techniques that can do this. This book will focus on the most popular type of do-it-yourself additive machine: the type that builds objects out of layers of plastics (technically, fused deposition modeling or FDM). However, there are other types of printers that are less common among do-it-yourselfers.

For example, there are machines that cut slices out of paper and stack them much as a plastic printer stacks plastic. Others use a laser to sinter metal powder or photosensitive polymer. The core idea, however, is the same: given a 3D model, use the computer to determine what each slice would look like if you cut it into thin slices and then build those slices. A typical printer might use slices of 200 or 300 microns. If you aren't used to thinking in microns, 200 microns is 0.2 millimeters. There are 25.4 millimeters in an inch, so 0.2 millimeters is about 0.008 inches.

Of course, an object created out of paper is going to be a lot different than one made of sintered metal and plastic parts will be different still. Even in plastic parts, the type of plastic can make a difference. You can create plastic parts that are strong enough to work as improvements for your printer, for example or repair household items.

There are things you can't do. Depending on your printer, there is some limit to the fine detail resolution you can achieve. Plastic melts with heat, so you can't make parts that will sustain high temperatures. Depending on the plastic you select, your parts might be sensitive to certain solvents, as well.

Although a lot of people print items like cups or shot glasses, this isn't a good idea. Only some plastics are food safe and even if you had food-grade plastic, it is unlikely that your printer is food safe. It isn't just a matter of cleanliness. Some plastics could release unsafe chemicals into your food. Other materials can allow organic material to seep beneath the surface of the material where it could go bad and cause illness later. Your printer also contains plastics and metals that could render a printed object unsuitable for food use.

But there are plenty of things you can make. Figures 1-1 to 1-5 show many items created with an inexpensive ($600) plastic printer: a clock, a vise for printed circuit boards, and several ear bud winders, for example.

Figure 1-1. Earbud holders printed at .1mm, .2mm, and .4mm layer heights

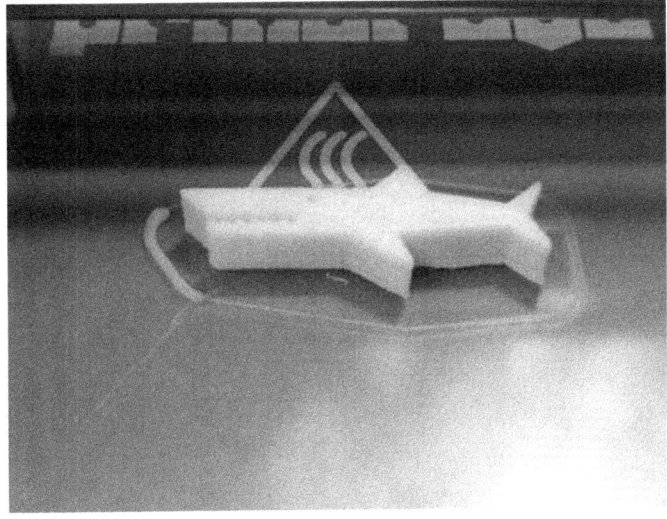

Figure 1-2. A chip bag clip in the shape of a shark fresh off the printer

Figure 1-3. Two pictures of one lithophane. The image on the left is not backlit. The right hand image shows the same lithophane when lit from the back. Thicker plastic shows up darker than thinner plastic forming the image.

Figure 1-4. Printed circuit board vise holding a (non-3D printed) printed circuit board

Figure 1-5. A clock including a (non-3D printed) clock mechanism

The steps to going from an idea to a printed object will vary a little depending on several factors, but, in general, you'll work with the following workflow.

1) Use a CAD program to create a model of a 3D object and save it as an STL file (STL stands for stereolithography—a different type of 3D printer, but the file format is very common). Some software directly will save to this format. Others will require an export step. There is also software that can convert between different formats that you could use. If you don't want to start with your own design, you might find an STL file on the Internet and use it.

2) Run a slicer program on the STL file. The slicer will take parameters about your printer, they type of plastic you are using, and other information you provide and use it to figure out how to slice the object in the STL file into dozens or hundreds of slices. The output of the slicer will be in a standard file format called G-code. This format is actually the same format used for traditional CNC machines, although there are a few extensions commonly used for 3D printing.

3) You'll make sure your printer is loaded with plastic and set correctly.

4) You'll use a piece of software to send the G-code to the printer using the USB or serial port. This software will usually show you some status information about the print job. If you don't want to tie up your computer, you can also (on some printers) save the G-code to a memory card and use the software to make the printer read the card and print the object without further computer intervention. A few printers have built-in screens and controls and don't need a computer at all to run the print job.

There can be, of course, some variation. The software you use in step 4, for example, may integrate a slicing program either directly or by providing a way to launch an external program to do the slicing.

You'll see more details about each step of this process in future chapters. First, however, you need to understand more about the printer and the material you'll use for printing.

In this chapter, you've learned about the types of 3D printers and what kind of parts you can (and can't) make with readily available printers. You should have a good idea of how a part goes from a computer design to a printed part. Now you just need to buy or build a printer!

2 MATERIALS

A car is no good without fuel and a 3D printer is not useful without some material to use to build your designs. Some printers need metal powder or polymer resins. Some make models out of thin slices of paper. The printers you'll read about in this book use different kinds of plastics. Most often, the printer will use plastic filament (think of something that looks like the line that a lawn trimmer uses). The current trend is to use filament that is 1.75mm in diameter but many printers still use 3mm filament.

You will usually buy filament wound on spools (see Figure 2.1) although small quantities might be "air core"—a fancy way of saying they are just coiled over nothing. Some commercial machines require you to buy their plastic in some sort of cartridge, but the printers you'll read about here simply take filament.

Figure 2.1 Spooled filament

Filament isn't very expensive—you can buy a kilogram (2.2 pounds) of certain kinds of plastic for US$25 or so. That much plastic will go quite a long way. Even so, plastic comes from the manufacturer in pellet form (paradoxically, they chop the pellets from filament) and the pellets are less expensive than filament with tightly controlled diameters. Recently, several machines have appeared that will create filament from pellets, but they are still relatively uncommon. However, this may be the start of a trend, and we may one day see printers that start with pellets instead of filament.

Type of Materials

Table 2.1 shows the types of filament in common use in today's printers. There are other possibilities (like polycarbonate) that need special care because of toxic fumes, so you don't see those as often. ABS is a good general purpose plastic. It is strong and reasonably heat resistant. However, it is subject to warping while printing (a subject you'll read about in Chapter 7). With care, however, ABS is a good choice and many people have never used any other material. ABS does give off a smell when you print. Some people find that it hardly smells, while others find the smell very offensive and develop headaches. Regardless of your sensitivity to the smell, you should definitely work in a well ventilated area regardless of the type of plastic you choose.

Type	Nominal Temperature	Notes
ABS	245C	Very common. Strong and reasonably heat resistant. Not sunlight resistant. Prone to warping during printing (requires heated bed).
PLA	220C	Very common. Made from corn starch and sugar. Better sunlight resistance and less warping (heated bed not required). Less resistant to heat.
Nylon	245C to 300C	Very tough and inexpensive. Requires ventilation and is difficult to print with because it tends to be "runny" and also is difficult to adhere to most print beds.
Polycarbonate	265C to 305C (or higher)	Very strong, low warp. Higher temperatures allow it to print clear. Requires good ventilation. Many 3D printers can't achieve the high temperatures required to get good prints with polycarbonate.
PVA	190C	Useful for printing supports that will be removed later. Water soluble and even humidity will degrade it.

Table 2-1 Common printing filaments

Another popular choice is PLA. This plastic is made from plant material and smells like waffles when it is hot (ABS has a "burning plastic" smell that people have a varying sensitivity to). However, it is not nearly as heat resistant as ABS. Many less expensive printers use PLA because it doesn't need a heated bed to prevent warping. However, there's no free lunch. PLA stays molten longer so you usually need some sort of fan to cool the printed object to get the best results.

Nylon is another common choice. It is chemically inactive and very strong. Most simple printers can't reach the optimal heat for this material, but can get close enough. Another issue with nylon is that it has a tendency to be very thin and runs out of the print head even when you don't want it to.

Many lawn trimmers use nylon string as a cutting element. Some people have experimented with using it instead of nylon filament. However, you should avoid this unless you have a good grasp of the chemistry involved. Many brands of trimmer nylon contain fiberglass to strengthen it. It could also have other materials and might release noxious fumes when heated. It might cost a little more to buy nylon made to print, but it is worth knowing that you have material that is safe to print with.

The final entry in Table 2.1 is not very useful by itself. PVA is the kind of plastic used in most white glues. It isn't a very good plastic for printing but it has one special property. It is water soluble. Sometimes, when printing with another material, you want to print something that would be in the middle of the air. Usually, you want to orient your design so this doesn't happen (a topic covered in Chapter 5). Sometimes that isn't possible and you will have a piece sticking out. The answer is to print support material under the part that is sticking out. This can be done automatically by software, but the problem is the material. If your printer can only print one kind of plastic at a time then the support material will use the same material. It can very difficult to remove. If your printer has two (or more) heads, you can use PVA for support material. Then, when the print is done, you can use water to dissolve the PVA completely.

If your printer doesn't support multiple heads, don't worry. You can print a lot of amazing things with no support—even things that have gaps. With a little care, you can use ABS or PLA for support material; it will just take a little more finishing after the part comes off the printer.

Exotic Materials

There are many exotic variations on plastic. For example, PLA is available that glows in the dark, and some that is flexible when printed. You can get filament that changes colors with temperature. That way you can print at one temperature and get one color and then raise the temperature slightly and get another color.

One very interesting material is Laywoo-d filament. This is a PLA that contains wood fibers. There is also Laybrick that looks like sandstone. Not only does the printed part resemble wood, but by controlling the temperature you can make the "wood" lighter or darker. With the right software, you can print a facsimile of wood grain. Because the filament contains actual wood fibers, you may have trouble printing it with a very fine print head. Usually, a smaller print head nozzle is better, but in this case, you can have too much of a good thing. Consult the filament vendor to determine what print head nozzle sizes they support.

The holy grail of printing filaments would be one that could conduct electricity. Sometimes you will find PLA that is conductive, but usually this has a very high bulk resistance. It is suitable for printing, for example, containers for electrostatic sensitive devices. They aren't usually useful for printing wires and other electronic components. However, there have been some experimental filaments that may be useful but, so far, they aren't widely available.

Some people experiment with printing polycarbonate and a few other kinds of plastics. However, these often require high temperatures and may have noxious fumes, so be careful before running something through your printer to make sure it is safe.

What Makes Good Filament?

There is a wide range of prices for filament and, as you might expect, a wide range of quality. There are several key parameters that determine how successful your prints will be:

- Roundness - Cheap filament can have an elliptical cross section which can cause problems when printing. If you measure the diameter of the filament with calipers and then turn the calipers 90 degrees, the reading should be very close to the same.
- Diameter variation - Another problem with cheap filament is that the diameter may vary at different points in the spool. It doesn't matter that the filament is exactly the right size. So a 3mm filament might really be 2.7mm or 3.1mm. That's not a big problem as long as you measure it and put the right size in your software (see Chapter 5). The problem is if the start of the spool is 3.1mm and the middle of the spool is 2.8mm.
- Contamination - Filament should be free of contaminants. If you print in an area that has a lot of debris (perhaps cat or dog fur) you might need to arrange for the filament to pass through a lint-free cloth on its way to the printer.
- Dryness - Plastic needs to be dry before printing. Usually, the

vendor will ship it in a sealed package with desiccant packages. If you leave the plastic out, you might dry it in an oven or under a heat lamp. However, most people place it in a reasonably air-tight container filled with silica gel (see Figure 2.2). One ready source of silica gel is crystal kitty litter. One bag will probably last you a very long time.

Figure 2.2 - A drying box

Making the Choice

You might see a good deal on plastic and be tempted to buy some before you buy your printer. Don't! Hang in there until you know what your selected printer needs. At the very least, you need the right diameter and possibly a material that your printer can handle. In general, you ought to start with PLA or ABS since these are widely available and easy to work with. Save the more exotic materials and variations until you have some experience under your belt.

3 THE HARDWARE

Although a 3D printer runs on software and CAD drawings, it is primarily a hardware project. Depending on the style of the printer, it could look like a kitchen appliance or -- more likely -- it will look like a mad scientist's latest invention, bristling with wires, motors, belts, gears, and shafts.

The Good News

Don't let the mechanics of a 3D printer scare you. A printer isn't much more complicated than, say, a bicycle. When I was shopping for my first printer, I was intimidated and wanted to buy a printer that was preassembled. I wound up with a kit, and in retrospect, I think it was a good idea. Not only did I learn a lot about the printer, but the state of the art today is such that, unless you spend a small fortune, you can count on having to fix your printer, adjust things, or unclog jams. If you've built the printer, you will be better prepared to tear it apart and rebuild it when the need arises.

Every printer is different, but there are common features that most printers share. This chapter will talk about the hardware that lets the printer deposit material where you need it to make your designs a reality.

What is a Printer?

Most printers fall into the category of Cartesian robots--a fancy word for a robot that can move in three dimensions. If you are looking at the front of a printer, it is conventional to consider the left to right direction to be X. The Y axis is from front to back. The vertical motion, of course, is the Z axis.

The majority of printers use some sort of motorized arrangement to move in two axes and then a separate arrangement moves in the third axis. For example, the popular Printrbot has a printhead that moves in the X direction. The entire carriage that the head moves on goes up and down (on the Z axis). A separate bed moves in the Y direction independently.

Other printers might choose different combinations, but the idea is the same. There are other options, but they are (so far) less common. For example, a Rostock printer (a type of polar robot) uses a head that "floats" on a triangular arrangement of three supports attached so they can independently move vertically. By controlling the length of each string, the print head can swing to any point in the three dimensional build space.

Even a Rostock printer, though, needs some way to move something. Just about every printer uses a special motor called a stepper motor (sometimes just called a stepper). You normally think of a motor as a device that spins when you apply power to it. A stepper is a bit different. By sending a special sequence of pulses to the motor it will move exactly a certain number of degrees. By changing the pulses you can make the shaft turn clockwise or counter clockwise. Figure 3-1 shows a typical stepper motor mounted to a printer.

Figure 3-1. The X-axis stepper motor on a 3D printer

Steppers come in many different configurations, but it isn't unusual to see stepper motors that turn 1.8 degrees per step. That means that it takes 200 steps to make the motor do a full revolution of its shaft. The precision of the stepper motor is key to getting good printing results. An ordinary motor lacks the precision unless you employ additional sensors to determine the position of the shaft, so they are not often used for 3D printers.

Many printers can also "microstep" motors. That means that it can move the motor (with someone less accuracy) to smaller microsteps in between the primary steps. A microstep of X8 or X16 is common. For a 200 step motor, that translates to 1600 or 3200 steps per revolution.

The physical size of a stepper motor (which usually correlates to the torque it can produce) is specified by a NEMA (National Electrical Manufacturer's Association) number. A NEMA 17 motor, for example, is not as large as a NEMA 23 motor. Unless you are building your own machine from scratch, your printer calls for a certain size motor and that's what you need.

Another differentiator of stepper motors is the way they are driven. A conventional motor has two wires, but a stepper motor will many wires that control the coils inside. Some steppers require the drive voltage to be reversed in some situations. These are called bipolar motors and usually have 4 or 8 wires. A unipolar motor uses a special arrangement of coils and usually has 5 or 6 wires. The electronics to drive a unipolar motor is simpler, but the produced torque is usually less when compared to a bipolar motor.

You may have noticed a problem. Most motors, stepper motors included, move a shaft around in a circle. To build a printer you need things to move linear (straight). There are several common ways this can occur.

One of the easiest ways to make a rotating motor move something in a straight line is through the use of a leadscrew (see Figure 3-2). The idea behind a leadscrew is simple. A leadscrew is just a rod with threads cut in it (just like the shaft of a very long bolt). If you spin the leadscrew (with a stepper motor, for example), a nut on the leadscrew (see Figure 3-2 and 3-3) will move if it is captured to where it can't spin. A wooden or metal platform that has nuts attached to it will ride up and down on the leadscrews as it spins.

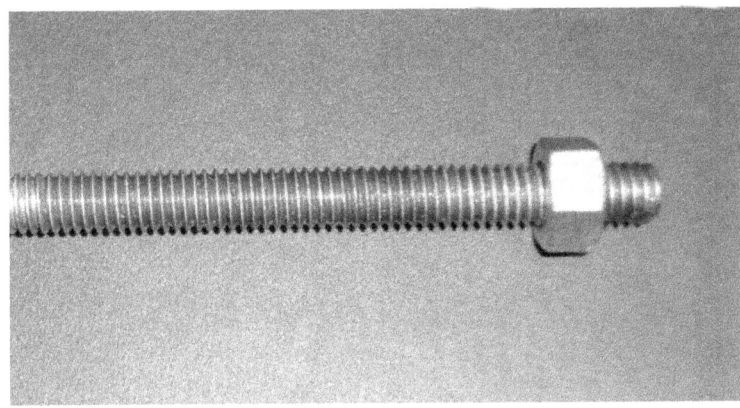

Figure 3-2. Threaded rod used as a leadscrew with a normal nut

It sounds simple, but it is a little more complicated to get it to work well. You want whatever is travelling on the leadscrews to keep most of its weight off the leadscrew itself. That's why you often have smooth rods in addition to a leadscrew. The smooth rod holds most of the weight.

A part that is used to transfer a load from one part (like a platform) to another part (like a rod), is called a bearing. A bearing can take many forms, but for this application, it will usually look like a metal tube (think of a socket from a mechanic's tool chest) that has little steel balls inside (ball bearings, of course). The platform will attach in some way to the bearing and the rod will slip through the bearing.

Instead of a smooth rod, some machines will use slides for a similar purpose. These slides are similar to the ones that you'd find on a drawer in a desk or filing cabinet. Their purpose is the same as the smooth rods: to bear the weight of the moving part, so the leadscrews (or other motion mechanism) doesn't have to labor against the weight.

Cheap machines will use common threaded rod (sometimes known as all thread) and drill rods. These can work but they are often not as precise as you would want for the best results. Special rods made for motion control are typically made to higher tolerances (and, of course, cost more, too). They will be straighter and have more precisely sized threads as well as threads per inch (or millimeter).

Motion control rods aren't just more precise. They usually use a special kind of thread known as ACME. These threads are designed to transfer power. Not only will an ACME thread transfer more of a motor's power to the moving part, but it also will have less wiggle and backlash. When you are printing you don't want anything wiggling. Backlash is a common problem with motion control systems. If you are moving in one direction and then change to another direction, it may take a little time for the reverse in motion to take effect. That's due to backlash.

ACME threads will have less backlash than conventional threads. In addition, you can use special nuts that minimize backlash. Delrin (a kind of plastic) nuts don't have as much backlash as metal nuts (see Figure 3-3). Some metal nuts use springs to minimize backlash. Speaking of Delrin, better systems use a softer material for the nuts on a leadscrew so that wear will occur on the nuts and not on the leadscrew or threaded rod. For example, a steel rod works well with a brass or Delrin nut, instead of a steel nut.

Figure 3-3. Special Delrin nut used with a leadscrew to translate rotation into linear motion

On the other hand, a 3D printer doesn't put a lot of wear on mechanics compared to say, milling away aluminum or steel. Many printers use inexpensive all thread and steel nuts and get good results. There's good and there's good enough. For many printers, an inexpensive leadscrew will do fine.

There is at least one more option for a motion control rod, although they are expensive and rarely seen on 3D printers: the ball screw. A ball screw is very similar to an ACME threaded rod but instead of a nut moving on a thread, the threads are traps for ball bearings and the moving part rides on the ball bearings. Some high end CNC machines use ball screws, but they are costly and probably not really necessary for a typical 3D printer.

You read earlier how a microstepping motor can make 1600 or 3200 steps per revolution. That, combined, with a leadscrew can create precise linear motion. Suppose you have an ACME leadscrew with 10 turns per inch. That means for every inch the captive nuts travel the motor would make 16000 or 32000 microsteps! If you really like English measurements, then your basic resolution could be as high as 1/32000 of an inch. Most 3D printing uses metric measurements, and with 25.4mm in an inch, the resolution is about .0008 mm per microstep.

An Alternative: Pulleys

There are a few problems with leadscrews. First, a good lead screw isn't cheap. In addition, the problem with that 1/32000 inch resolution is that it makes a machine relatively slow. That isn't to say you don't want a machine that uses leadscrews. On the contrary, high end machines that mill metals usually use leadscrews. However, for a plastic 3D printer, you don't have to use leadscrews. Many machines (like a reprap Wallace, for example) use a leadscrew for one axis, but use a different method for the other two axes.

Another common scheme is to use a pulley (usually a toothed pulley) to move a belt. The belt is attached to the moving part. This is inexpensive and gives good results if done properly. The downsides are that it does take some adjustments, is subject to wear, and takes some extra space.

Think of a platform with a belt. The platform rides on two smooth rods and bearings. To make it move, there is a belt under the bed and mates with a stepper motor and a pulley. The belt will have to be tight so as not to slip, but assuming there's no slip this is a good way to get reasonably fast motion.

Not many machines use smooth pulleys, and there are several toothed pulleys to choose from. Many inexpensive printers use timing belts (such as XL belts) and pulleys. The problem is, a timing belt is made to move in one direction. Because that's their intended use, the belt may not have good performance when it changes direction. There are more exotic belts (like a GT2 belt) that are made for motion control, and better machines will use these. Examples of both types of belts appear in Figure 3-4.

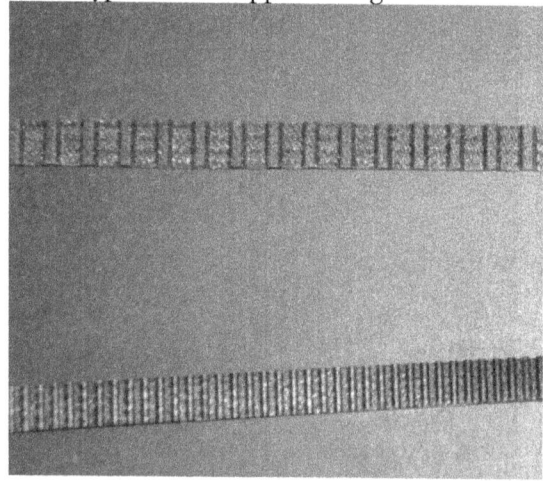

Figure 3-4. XL belt (top) and GT2 belt (bottom)

One potential problem with belts is that they can stretch when kept tight. Some machines have experimented with using high-grade fishing line instead of belts with good results (obviously, these would not use toothed pulleys). Another option would be to use a gear with a rack (a linear gear) although this is relatively rare.

Toothed gears have a certain number of teeth and the belts have a pitch (number of teeth per inch or millimeter). You can do the math to figure out how many millimeters a full turn of the pulley will move something attached to the belt.

For example, a belt with a 2mm pitch and 10 tooth gear means that every full turn will move 20mm. If the stepper motor has 3200 microsteps, then each step is .00625mm. Therefore, 160 steps will make the belt travel one millimeter.

Other Bearings

The bearings that look like a mechanic's socket are linear bearings (see Figure 3-5). They will have designators like LM8UU. The number 8, in this case, means 8mm. The UU (or other letters) tell you about the size and construction of the bearing. These bearings let a shaft move in a straight line.

Figure 3-5. An 8mm linear bearing

There are other kinds of bearings you may encounter in a printer. A skate bearing (Figure 3-6), for example, looks like a doughnut and is commonly used in roller skate wheels. As you might expect, these deal with rotating shafts. For a printer, these bearings may be used as pulleys or other rotating wheels to move filament.

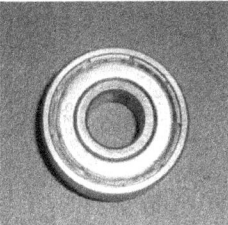

Figure 3-6. A skate bearing

Another type of bearing you might encounter is a pillow block. A pillow block is a bearing fit in a holder that attaches a shaft to a flat object.

Technically, the block is just the holder, but people commonly call the whole assembly including the bearing a pillow block.

When using bearings, it is important not to apply too much force to make a shaft go through the device. There are tiny ball bearings inside that you can easily knock out. A bearing that feels gritty or sandy when moving probably lost some of its internal bearings. It might also need lubrication.

Extruders

There is a kind of fourth dimension required for a 3D printer. Somehow, you have to move the plastic filament into the hot printhead. There are many ways to do this, but, in general, the extruder will use some sort of tension device to grip the filament, and a motor with some mechanism to move the filament. Most extruders today use stepper motors, although there have been some designs with other types of motors.

A conventional extruder will ride on the print bed (or hot end). This is easy to make work, but it adds weight to the moving print head, which can limit speed and accuracy. It also makes it hard to put multiple extruders over the print area.

To combat these issues, some printers use an arrangement called a Bowden extruder. This is very similar to a conventional extruder except the filament is pushed into a tube (usually made of PTFE--more commonly called Teflon). The tube carries the filament to the hot end. This keeps the moving part very lightweight. It also means the bulk of the extruder isn't next to the hot end, so it is easier to have multiple hot ends and extruders for making multiple colors or support material (see Chapter 1 for more about multiple extruder machines).

Enclosures

Some printers are enclosed in a box while others are not. All things being equal, an enclosure can help the machine stay at the set temperature and keep down dust and contamination. On the other hand, many printers don't have an enclosure and work just fine. Of course, you can always build your own enclosure if you decide you want one.

Tools You'll Need

There are a few tools you will probably want to use with a 3D printer, some common and some a little more exotic:

- A set of calipers, preferably digital - From calibrating your machine to measuring printed parts or parts you want to mate with or duplicate, you'll find a caliper to be one of your most-used tools.

- A spatula style-paint scraper or putty knife - It can be difficult removing a printed part from the printer bed and these can help.
- Needle files - You can expect to have to clean up parts, on occasion. You may also want some fine emery cloth or sandpaper.
- Fine diagonal cutters - In addition to wire, you may have to snip stray bits of plastic away on finished parts.
- Needle nose pliers - Sometimes you need to snatch a stray piece of hot plastic off a hot bed. You won't want to use your fingers!

You'll also need a good variety of the usual tools. You'll need screwdrivers, hex keys, and possibly sockets or wrenches, especially when building your machine. Keep in mind that many printers are primarily metric and even the ones that aren't, typically have some metric parts (for example, the stepper motor mounts). Metric drill bits are notoriously hard to find in the US, so if you need them (and you may not) plan on ordering them over the Internet.

In addition, you might not have to have these tools, but they can be useful:

- A small level - Getting a printer actually level isn't that important. What you really want is the pieces of the printer to be level with respect to each other. However, a small bubble level--or even better, a bullet level--can be helpful.
- Soldering iron - Many printer kits come with assembled electronics, and--obviously--a printer that isn't a kit won't need one of these, at least, at first. However, you will eventually need to fix something or will want to make an enhancement. Be sure to use rosin core solder and keep in mind that bigger is not always better when trying to solder fine wires and components.
- Digital Volt Meter - These are indispensable for troubleshooting and general checkout. They are inexpensive and widely available, so if you don't have one, plan on getting one.

Of course, you might need other tools for your specific printer, but the above lists will give you an idea of what tools you might need.

The mechanics of the printer are crucial to get a good print. Things should move exactly where you tell them to move. Nothing should move without being commanded to move. Things like backlash and imprecision in mechanics can ruin your prints.

However, the mechanics is only part of the story. The mechanics don't do anything without electronics (the topic of the next Chapter) and software (Chapter 5). However, you can't build a great house on a poor foundation and you can't get a great printer with bad mechanical components.

4 ELECTRONICS

In the last chapter, you found out about the hardware used by many 3D printers. The mechanics are important, of course, but without some electronics, the mechanics won't do much.

Connecting

Most printers connect to a PC via a USB port. Some older printers use a serial port and even the USB devices look like a serial port to your software, which expect a serial port. In addition to the computer connection, some printers will have a memory card interface. You still may need a computer to start the print job even with a memory card. Some printers have an LCD screen and buttons so they don't need a computer at all.

You might wonder why a printer that doesn't have a control panel would use a memory card. Of course, you don't have to use the memory card—you can always drive the printer from the computer. But a long print job could tie up your computer for hours. By putting your design on a memory card and only using the computer to kick the job off allows you to disconnect or turn off your computer without interrupting your print. This is also a good idea if your computer is prone to crashing. Not all printers support a memory card, and many people don't use them even if they are available.

Something has to take input from the computer or memory card and translate it into motion of the motors and the print head. There will be one or more controller boards that handle this function. Many new printers use an all-in-one board that does everything. Some printers use older boards that will have one computer interface board and one or more motor control boards. The controller also has to manage things like heaters and sensors.

This controller board (or one of the controller boards, if the printer has multiples) will communicate with the PC or the memory card. You'll read more about what the computer sends to the printer in Chapter 5. Typical printers accept commands using a text format known as G Code. These commands will make the printer move the head to different X, Y, and Z coordinates, extrude plastic, and report status back to the PC. A few printers use a similar format but using a compressed binary form, but the idea is the same.

Power

The controller itself needs a little power, but it also has to control a lot of power that goes to motors and heaters. For that reason, your printer will have a hefty power supply. Some printers use a common PC power supply which is not especially attractive, but has the advantage of being cheap and available. One disadvantage of a PC power supply is that the highest voltage available is 12V. Because of the high power consumption, some printers use a higher voltage (which leads to less current in the wiring for a given level of power). Your printer should come with some kind of power supply, but don't expect it to be wimpy wall transformer!

Heaters

What's taking all that power? The stepper motors, of course, draw quite a bit of current, at least in their peak consumption. However, keep in mind that stepper motors run on pulses and some of the motors don't run all the time. Another big consumer is the heater that melts the plastics which runs during most or all of the printing process. Many printers also have a heated bed to aid in having plastic stick to the bed and not warp (see Figure 4-1). That heater will also take a lot of power.

Figure 4-1. The red surface is the heated bed covered in polyimide; the red nozzle is the hot end

The temperature of the bed is not terribly critical, so the controller will probably just turn it on and off to roughly control the temperature. A sensor—a thermistor, probably—tells the controller what the current temperature of the bed and the PC software will display the temperature. In general, if you are going to print ABS, you need a heated bed for good results. Other materials may not need a heated bed although you still may find one useful. Many inexpensive printers use PLA plastic and omit the heated bed to save cost and relax requirements on the power supply.

Heated beds are often made from a printed circuit board or using polyimide film heaters (sometimes known by the trade name Kapton). Electrically, these heaters look like big resistors and will take a lot of current. Since the bed area is fairly large, it may help to put some sort of insulation under the bed. Be sure the material you use can stand the heat. Reflectix is a popular option. Cardboard or cork can be used, as well as baking silicone.

The hot end temperature is more critical since the plastic will have different qualities depending on the tip temperature. In fact, some plastics change color depending on the temperature. The tip of the hot end is also small compared to the bed so, the temperature tends to vary more. Most controllers will use a more sophisticated algorithm such as a PID (proportional integral derivative) controller. Hot ends are often insulated with braided silicone/fiberglass sleeving to both stabilize the temperature and help prevent accidental contact with fingers.

There are several common hot end designs, but all use some kind of heater to melt the plastic as the extruder feeds it. Some hot ends use a resistor, others use a nichrome wire element or a ceramic heater cartridge. In any case, it takes a significant amount of power to heat the plastic to the melting point which, for ABS could be as high a 240C. Keep in mind that 100C is the temperature of boiling water!

Fans

What else does the controller have to manage? Some printers have fans either to cool motors or to cool the filament as it comes out of the hot end. The motor fans may run all the time or may be temperature controlled. If a fan cools the filament, the software will control it (you'll find out more about that in Chapter 5). Certain plastics (notably PLA) don't solidify quickly and will benefit most from a fan. However, with any material, a fan can help when printing overhangs, small features, and bridges (a bridge is formed when, for example, the printer is printing the closed end of a vertical slot and joining two pieces of plastic that have air between them). You can see a fan in Figure 4-1.

Sensors

The final task of the controller is to monitor various sensors on the printer. As mentioned earlier, heaters usually have a temperature sensor. Keep in mind that the temperature sensors have to be the type the controller expects (or, you will have to recalibrate the software). In addition, the way the sensors are mounted will affect their accuracy. It is always a good idea to check the temperature will a real instrument or judge the results and remember that the number reported by the printer may not be really accurate.

Printers will also have a number of other sensors. Most commonly there will be a mechanical, optical, or magnetic switch that allows the controller to determine when an axis is in a known position (known as the home position). For example, a printer might have three microswitches, one for each axis. On startup, the controller will drive the motors on each axis until the switch trips. That way it knows the exact position on that axis. Once all three are set to home, the controller can determine where the head is by counting the number of steps (or microsteps) that it sends to the motors. The printer may have an adjustment on one or more of the axes home switches so you can set the exact home position. You'll see in Chapter 6 how either you or the controller will probably need to adjust the Z axis, at the least, to ensure the plastic will adhere properly to the print bed.

Typically, the home sensors will show as a short circuit until the corresponding axis reaches the home position. This prevents the printer from continuing to drive a motor if a wire breaks loose. Note that although many people refer to these sensors as limit switches, they often are not limit switches in the truest sense of the word. The printer probably only reads them on start up. Once it knows the home position, it won't look at the switches again until the next time the software commands the printer to return to the home position.

Controller Programming

If you are so inclined, you can actually program the controller yourself. Most of the popular boards have some kind of bootloader (often based on an Arduino). You can usually (but not always) get the source code for the controller, modify it, and download your changes into the controller.

Of course, that's not for everyone. If you are experienced with microcontrollers, you can do almost any kind of customization to your printer you want. However, if you are just starting out, you might not want to experiment on your new printer.

The good news is you should never have to do this unless you want to try it. Even if there is a new firmware upgrade for your controller, you won't have to do any programming. Usually there will be an upgrade file and simple instructions for how to upload the new firmware to the controller.

Extras

There are other electrical devices that may be on a printer—lights, additional position sensors, or even cameras. However, this chapter covers the working electronic parts of a common printer. With the hardware and the electronics in hand, you need two more things: some software and some design to print. The software is the topic of the next chapter.

Electronics and hardware give you the physical tools to print. But just as a computer won't do anything without an operating system, a 3D printer needs software. Lots of it! It also needs models to print. Those are the topics of future chapters.

5 THE SOFTWARE

Once you have the mechanics and the electronics in place, your printer still needs some software to do its job. In fact, it needs lots of software.

The details will vary, but Figure 5-1 shows a typical software workflow for taking an idea and converting it to a solid object. This chapter will focus mainly on starting with an STL file (the second row in the figure). You can download STL files from the Internet for a wide variety of printable objects. If you want to do your own designs, you'll want to investigate a CAD program. Tinkercad is simple to learn and runs in your browser. If you are a programmer, you might want to read Appendix 2, which has an overview of programmatically building STL files using OpenSCAD.

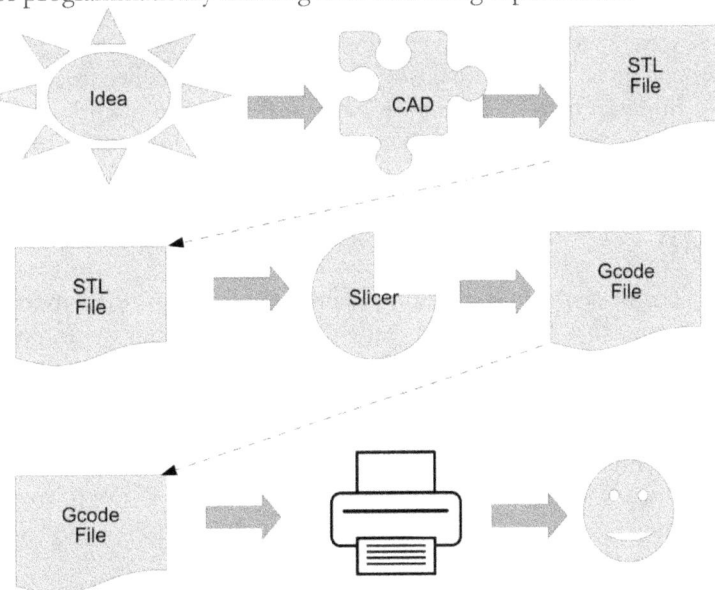

Figure 5-1. Workflow from idea to printed part

The printer actually has two pieces of software. There is a host program on your desktop computer. There is also firmware that runs on the board inside the printer. The printer firmware is really part of the hardware. You should never need to change it unless you are an embedded programmer and want to try new things, but that's outside the scope of this book. For ordinary printing, the firmware on the printer should be sufficient and if there is ever an upgrade to the firmware, you shouldn't need to understand it any more than you understand a device driver update on your PC. You just load it and go.

So ignoring the firmware, the first piece of software you'll encounter is the slicer. This takes a 3D model file and generates G-code which is the standard way to control machine tools, including 3D printers. The slicer will have a lot of options. Some of these tell it about your printer (for example, how big the bed is). Other options let you specify how you want the model to look (for example, the height of each layer or the interior density of the "fill").

The printer's firmware communicates with a host PC program that usually has several functions:

Allows manual control of the printer so you can move the heads, heat up heaters, etc.

Monitors the printer's sensors and displays items like temperature

Sends G-code to the printer

These interface programs often include a way to visualize what the printer is doing and a way to call the slicing program. With some interface software, you can pretend you are doing all your work in the interface, even though it calls the external slicer program for you as needed.

In fact, there's no reason why software can't work as "all-in-one" although the current trend is to use separate programs for slicing and printer communications.

CAD

The next obvious question is: where do you get the 3D model file from? One answer is to download ready-made models from the Internet. Perhaps the most popular site is called Thingiverse (http://www.thingiverse.com). In fact, some of the models there can be customized so you can alter the models in predefined ways.

Eventually, though, you will want to create your own models. To do this, you will need to master some sort of 3D CAD (Computer Aided Drafting) program. Sketchup is one popular option, but there are many others, including some online options (see below).

Programs like Sketchup are reminiscent of a paint program but with three dimensions. That is, you use tools to draw objects on the screen, move them, and change their properties all in a visual manner.

For some projects, though, this is almost a hindrance. If you want to line up two holes and make them exactly 10 cm apart, you have to carefully move the mouse and count grid lines. Most tools have some way that you can simply type in coordinates to position things where you like.

The next logical step is to build objects using only typed in coordinates. That's the idea behind parametric modeling programs like OpenSCAD. Using these tools makes 3D design more like programming. You provide a text file that describes your object in terms of primitive shapes. Usually, you'll write the "program" in the program's built in text editor so you can render it as you go and check on your progress.

You might wonder how you can build a complex object out of basic shapes. The trick is the program can add and subtract shapes. So to make a panel with holes in it for mounting lights and switches you'd start with a cubic shape (the panel) and then create cylinders for the various holes. Then you tell the program to cut the cylinders out of the panel. You can also join objects together and make other combination types.

There are many CAD programs. Figure 5-2 shows a simple object created with OpenSCAD (actually, it is a single object printed four times). The "program" that created it appears here:

```
/* Really simple "mending plate" with two #6 holes
    -- I use this to clamp a GT2 belt,

      but it probably could be used
      for anything. Al Williams, April 2013 */
// Length of plate (mm)
length=17; // [10:100]
// Width of plate (mm)
width=6; // [5:100]
// Thickness of plate (mm)
thickness=3; // [.5:20]
// Offset from edge/hole spacing (mm)
edgeoffset=3.5  ;  // [1:10]
// Number of holes (should be even)
n=2; // [[1:100]
loopct=n/2;
difference() {
cube([length,width,thickness]);
for (i=[1:loopct]) {
  translate([edgeoffset*i,width/2,1])

    cylinder(h=thickness*10,r=2,center=true,$fn=100);
  translate([length-edgeoffset*i,width/2,1])

    cylinder(h=thickness*10,r=2,center=true,$fn=100);
  }
}
```

Figure 5-2. OpenSCAD-generated plates

You can read more about OpenSCAD in Appendix II. If programming isn't your style, you might want to try an online CAD program. Most of these are easy to use, work without installing anything, and often have tutorials to help you get started. These tend to come and go, but here's a few to try:

- http://www.tinkercad.com – Possibly the best and since it was bought by AutoDesk, should be around. Great tutorials.
- http://www.3dtin.com – Another browser-based modeling program. To make holes, you need to use the "make your own geometry" command.
- http://shapesmith.net/ - A cross between a graphical approach and parametric. Well done, but a little harder to learn.

Most CAD software can import an existing model and modify it. For example, my binary clock design (see Figure 5-3) started life as a normal clock (see http://www.thingiverse.com/thing:39899). I simply used a CAD program to remove the original numbers and replace them with my own.

Figure 5-3. Binary clock (files at
http://www.thingiverse.com/thing:113334)

If you don't want to use CAD software in your browser, you can download many free CAD programs. Google SketchUp is popular, as is FreeCAD. There are many options if you are willing to pay for your software, depending on the type of computer you want to use.

Besides CAD software, you can also use 3D modeling programs if you are artistic enough. Blender is popular for this type of work. Art of Illusion is another program that may be a little easier to learn. These programs are better for modeling what you would think of as a sculpture and perhaps not as easy to use if you want to make parts with precise measurements and geometric shapes.

Slicing

Figure 5-4 shows a common slicing program known as Slic3r. It has several features:

- It can build up a "plate" of one or models with different positions and scalings
- It can produce G-code for the plate you wish to print
- It can output slices to SVG files
- You can give it separate configurations for different types of filament, different printers, and different settings

To use Slic3r, you first have to set up many parameters for the print. For example, you will want to pick a layer thickness (that can't exceed your nozzle diameter) and print speeds. These go in the Print Settings tab. The Filament Settings tab lets you tell the software the diameter of your plastic and the temperature it should melt at. The Printer Settings tab had information about your specific printer.

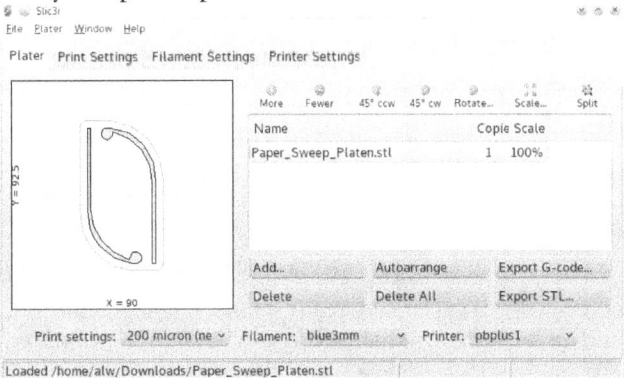

Figure 5-4. Slic3r, a popular slicing program

You can save all these configurations, so you generally only have to set things up once. In the Figure, the settings (near the bottom of the screen) are "200 micron (new)" "blue3mm" and "pcplus1."

After you've set these up once, you use the Add button to read one or more STL model files. They will appear on the "plate" to the left. You can rearrange them, move them, and even scale them. When you are satisfied, you press "Export G-code" and the resulting file is ready to send to the printer.

Some print control software automatically calls a slicing program for you, at which case you might never see the actual slicing program. For example, Repetier Host can call to Slic3r or Skeinforge (another slicing program) and you generally don't have to open them yourself to use Repetier Host.

Slic3r is not the only choice. Skeinforge is another popular choice, although it is somewhat slow and has a bewildering set of options. Meshlab, Cura, Repsnapper, netfabb Studio, and Kisslicer are all capable slicers, as well.

Before you can slice, you will have to tell the slicing software something about your setup. In Slic3r, you will want to set your bed size, bed center, and other important details in the Printer Settings tab. You can also set up Custom Gcode in the same tab. You may or may not want to do this. Most printers require some setup before starting, and you probably want the print head to move to a neutral position when it is complete. One way to do this, is to set custom Gcode in Slic3r. There are four boxes to enter custom code: one for startup, one for ending, one for layer changes, and one for tool changes.

If you enter the custom code in Slic3r, though, your Gcode files will be set to whatever parameters you select. For example, a common use of a startup code is to set the calibration parameters for the printer. If you put them in the startup code for Slic3r then any time you change the calibration you will have to reslice before printing. That's not a big deal, and for some print software, it may be your only option.

Some print software (for example, Repetier Host, discussed below) allow you to enter custom Gcode. In this case, the print software will manage sending the required commands before (and after) processing the Gcode file. This is handy because changing things like calibration constants doesn't require reslicing the model. Of course, a Gcode file is just a text file, so you could theoretically open the Gcode and manually change things without reslicing. That's not terribly handy, though.

Another important parameter you have to set in Printer Settings is the nozzle diameter. If you have more than one extruder, you'll have to set it for each extruder. The slicing software is actually very sophisticated. Given a particular layer height, it knows how much volume of plastic it needs to extrude to fill a particular part of your model. It also knows the diameter of the filament and the diameter of the nozzle. It can use all of this to compute how much plastic has to go in to ensure the right amount of plastic goes out. It also accounts for the speed the print head is travelling and adjusts the feed of plastic in to ensure the right amount of plastic leaves the nozzle at the right time.

All of those factors play together: the size of the filament, the nozzle diameter, the layer height, and the head speed. The slicer controls the head speed and the layer height, but it is up to you to accurately tell the software how big the filament and the nozzle are. In addition, your extruder calibration will convert the slicer's commands (in millimeters) into actual motor steps, so if the calibration is off, your prints will not be good.

If you don't have enough plastic flowing, you will wind up with gaps and voids in your printed output. If too much is flowing, your prints might look good, but your inside dimensions will be off. If you flow too much out, it will pile up and the head will bump into the excess plastic on the next layer.

Other parameters regarding the printer will determine if the slicer commands the printer to reverse the extruder before it moves without printing. This is usually a good idea, because it prevents ooze which leaves strings when the print head moves from one spot to another without ejecting plastic. A good value here might be 0.5mm, for example. When the printer starts in the new spot, the command stream will cause it to force out the same amount of plastic to restart the stream.

You can also direct the slicer to command the print head to raise up a bit before moving. You shouldn't have to do this, but it sometimes helps if your bed is not perfectly level or your Z axis or E calibration is a bit off (it stops the head from dragging on previous layers that are higher than they should be). A word of warning, though. Enabling this feature will significantly increase the wear on your Z axis. Be sure to keep threaded rods lubricated or belts tight (depending on how your Z axis works) when using this feature.

Part of the equation, as mentioned above, is the size of the filament. You set this in Slic3r's Filament Settings tab (other slicers will have similar settings). You can usually get away with specifying the nominal diameter. For most printers that is 3mm or 1.75mm. However, if you measure your filament, you will find it probably isn't exactly that size. For more precise printing you should measure the filament with calipers at different spots and enter the average size. Obviously, the more consistent the filament size, the better your results will be. Don't assume just because your filament is 3.1mm at the start of a spool that will stay that way always. It isn't uncommon for a spool to vary across its entire length.

Another issue with some filament is that it isn't exactly circular. That is, measuring it one way will yield one result and turning the filament 90 degrees and measuring at the same spot will give another result. This is an indication the filament is elliptical in cross section. Ideally, you shouldn't use elliptical filament, but if you must, you might consider averaging both dimensions and then take an average of the averages at several points.

The Filament setting is also where you set the temperature for both the extruder and the bed. Some printers don't have heated beds, although for printing with ABS it is practically essential. If you use PLA, though, you don't need to heat the bed and printers that don't have heated beds typically use PLA for this reason.

Slic3r provides separate settings for the first layer on both temperatures. This is the first of many places you will see that Slic3r tries to help you get a good first layer. You'll see in Chapter 7 that the first layer is crucial to getting a good print. The best house will fail without a solid foundation and that's exactly the purpose the first layer serves.

The first layer needs to stick to the bed. This will do several things. First, it will keep the part from moving around on the bed. Second, it will keep the cooling plastic from warping, which will ruin the part as the head starts to bump into the upturned edges. Finally, it will provide a perfect surface for the next layer to attach itself to.

There are several items that contribute to getting good bed adhesion: layer height, speed, print surface, Z height, bed leveling, and temperature. It isn't unusual to increase the temperature of extrusion or the bed temperature to attempt to get a better adhesion to the first layer.

With Slic3r, the filament settings also contain the cooling options. If your printer has a fan, Slic3r can automatically turn the fan on when it thinks the print needs to be cooled. There are several reasons this occurs. First, if you are printing a very small layer, the plastic you lay down on layer 5 might not be solid by the time you start printing layer 6. That won't work well. Slic3r can slow small prints down to help fight this. It can also kick a fan on at a variable speed.

PLA stays molten longer than ABS so a fan is really useful with PLA. However, it can be useful with all kinds of plastic. It also helps with bridges: when you are extruding plastic over a gap. For example, if you were printing a letter "U" upside down on the print bed, the horizontal part of the U would be a bridge. The plastic needs to cool fast to prevent it from drooping into the gap. Of course, it will droop, but you want minimize the effect. Of course, that's just an example. In real life, don't print the "U" upside down, but in some cases you can't help making a bridge and a fan will help.

You don't usually want to turn a fan on all the time. The fan will cool the bed and the extruder as well as the plastic and that can lead to other problems. You want to run the fan as slow as you can and only when necessary. Slic3r and other slicers usually can control this fairly well with just a little input from you.

This all seems daunting, but you usually only need to set the printer and filament settings once and then you can mostly forget them. You might want different settings for different kinds of filament, especially if you use different types (that is, if you switch between ABS, PLA, nylon, etc.). However, Slic3r will save your settings with a name, so once you have your different settings done you shouldn't have to change them much.

That's not true of the first tab "Print Settings" (in Slic3r, at least). Here you set many things you might change frequently. The first of these is the layer height. Your part will be made of many thin slices of plastic, and this parameter sets how thick those layers are. A typical printer might use .3mm as the default layer height. Some printers can go to .1mm or even lower. The thinner the layers, the longer a print will take (because it will take more layers). However, thicker layers won't look as good.

A few rules about the layer height are in order. First, you can't set a height larger than your nozzle size. Smaller nozzles give better resolution, but they also cut you off from thicker layer heights. In general, the thicker the first layer is, the less critical alignment of your print head will be to get plastic to stick to the bed. Slic3r and other similar software will allow you to set a different height for the first layer (but it still has to be no bigger than the nozzle size).

If you really want to get picky, you should make sure your layer height is an even number of steps or microsteps of your Z axis system. For example, if you are using a threaded rod on your Z axis that makes 10 turns per inch, then each turn is worth 0.1 inch. Suppose that a full turn of your steppers requires 3200 microsteps. That means each step turns the Z axis 0.00008 mm (actually, just a hair less). That's 0.1inch divided by 3200 and converted to millimeters (times 25.4). As a result, you'd prefer to pick layer heights that are divisible by that number. In this case 0.16mm would be a better layer height than 0.1 or 0.2. At the very least, you'd like to set the height to be evenly divisible by a microstep.

However, to get started, you don't need to worry about it. Just pick a starting height as recommended by your printer manufacturer and you'll get nice prints. But when you are striving for the best prints, compute your layer height in terms of your Z axis.

To save time and plastic, most of the time you will print items that are not totally solid. The slicing software will want to know how many solid layers to put down for outside surfaces. In Slic3r, this is called the vertical and horizontal shells. Solid layers take longer to print and they also increase the chance of warping. Usually three layers on the outside will be enough unless you are using very thin layers.

What happens to the inside of the part? That's the infill. You can pick a fill density (for example, 0.2). This is what percent of the interior will fill with plastic (in the example, 20%). You can usually pick a pattern to use as a fill. Rectilinear is just straight lines, while honeycomb make hexagonal cells which are strong. If you are printing something with a circular cross section, you might want to use concentric fill as a pattern. If you want to print a hollow object, you can set the infill to zero.

Figure 5-5. Example honeycomb fill

Figure 5-5 Shows an example of a honeycomb fill. The part was stopped midway to show how the infill appears inside the part.

Another key item configured here is the print speed. In general, the faster the speed the less the quality of the finished product. Of course, at some extremely slow speed, the hot extruder may bother the plastic, but that would have to be very slow. How fast your printer can go at maximum depends on your hardware and how well it is all tightened. However, it is common to specify slower speeds around edges and on solid surfaces. The infill won't be seen, so going faster there is no problem. Another place to go faster is when the printer is creating a bridge (plastic over a gap). You need to get the plastic to the other edge of the gap fast before it has a chance to droop too far.

Slow speeds are good for outside surfaces. The first layer is an outside surface, and another case where you might want to go even slower to help with the all important bed adhesion.

Another feature that helps with bed adhesion is rafting, and brims. Slic3r doesn't support rafting, but it involves putting a layer of plastic (perhaps not completely filled in) down first that is meant to be cut away. Then the part prints on this sacrificial layer.

Slic3r takes a different approach. You can ask it to print a brim and it will draw a layer of plastic around the perimeter of your part. This can be useful with small parts that don't have a lot of area to hold on to the bed. You can control how wide the brim is and the idea is that after you remove the part you will peel or cut the brim away. When using colored plastic, removing the brim may leave imperfections in the colored surface, however, so for some plastics, you won't want to use a brim unless it is the only way to get the part to stick.

The other major settings (for Slic3r, at least) are the ones for support material. Suppose you wanted to print a letter "T" in the correct orientation. This would be difficult because the horizontal part of the T would hang in the air. With nothing to support the plastic, it would just fall down. You can get away with some amount of overhang because, with thin layers, each layer that forms the angle will rest somewhat on the previous layer. But once there isn't enough overlap (as in the "T") the print will fail.

If you tell Slic3r to generate support material, it will detect overhangs like this and print plastic in a pattern to support the overhangs. Like a brim, you are meant to remove the support material when the print is complete. It can be difficult to cleanly remove the support material if it is the same plastic you are using to print with (that is, you only have one extruder). If you have multiple extruders, you can use a water soluble plastic to generate support material and Slic3r will handle that.

Of course, you can draw support material directly into your CAD design, if you like. A smarter idea is to flip the "T" so that it prints upside down. However, with complex shapes, that isn't always possible.

The Print settings is one place where you will want to create multiple sets of settings. I have a setting for my normal layer height, and one for a draft height. I have one for solid parts, and one for hollow parts. Another enables support material for the cases I think I need it.

If you do it right, you shouldn't need to change the settings often, but you will frequently select a different set depending on what you are printing. Something round might get the concentric fill option while a statue might get a set of options that sets for hollow infill.

Print Software

Once you have a Gcode file from the slicer, you need a way to get it to the printer. Nearly all printers have either a serial port or (more likely) a USB connection that looks like a serial port to software. In cases where there is a real serial port or where the USB connection is just a USB to serial adapter, there is a real risk of the printer requiring data faster than the computer can send it. Boards with real USB connections usually don't have this problem (for example, the Teensylu or the Printrbot boards).

One popular host software is Repetier Host (see Figure 5-6). It can drive most popular printers, but if you are using the Repetier firmware on your printer, it can compress data before sending to mitigate data rate problems. However, most modern printers should have true USB connections and not have this issue. Even so, Repetier Host is a great program to use to drive your printer.

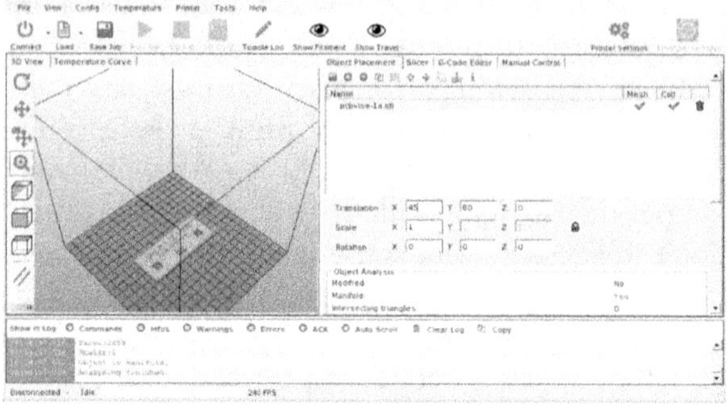

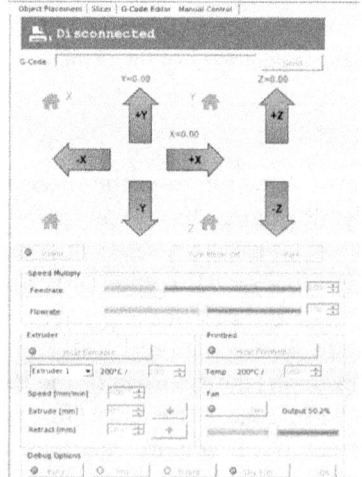

Figure 5-6. Repetier Host

Repetier is not the only software available, though. Many people started with Pronterface (also known as Printrun), a set of Python programs.

Basic printer software will let you load a file to print, manually move the printer mechanics, heat the heaters, and monitor the printer's status. More sophisticated printer software integrates directly with a slicer so that it appears you just load an STL file and print it. In reality, the printer software is still calling a slicing program to generate the Gcode. It simply hides that from you. Repetier Host (and other printer controllers) can even scale, copy, and rotate STL files before sending them to the slicer. This can be very handy to orient a model so that it prints well or to make multiple copies. You can see Repetier's screen for manually controlling the printer in Figure 5-7.

Figure 5-7. Manual control screen

Because the host software can control the printer, you'll need to enter some of the same setup (once) that you did to the slicing software. For example, the host software will want to know how big the bed is so it can move to either end (or the center). You should only have to set up your printer once and, unlike the slicing software, the host program won't care about things like the nozzle size or the layer height. It will only be worried about the motion of the mechanics.

You may note the two sliders in Figure 5-7 labeled Feedrate and Flowrate. The Feedrate slider allows you to adjust the speed of the mechanics as a percentage of the speed set by the slicing software. Normally, you'd leave this at 100 and the printer will obey the speed set by the slicing software. However, you can adjust it speed up or slow down all speeds by the same amount. So at 110 you'd have an increase in speed of 10%. Just keep in mind that going you should set the speeds you want in the slicing program and only use the slider for experimenting or special cases.

The other slider does the same thing, but affects the speed of the extruder. Increasing this is like changing your extruder calibration to get more plastic out. Reducing it will reduce the amount of plastic extruded for a given print volume. Again, you should really get your filament size, nozzle size, and extruder calibration set correctly, but this slider can be used to experiment and tweak a little.

Some printers have an SD card slot. If the printer has a control panel, you may be able to save Gcode to an SD card and just directly print. This is handy if you don't have the printer near a computer or if you have a computer that doesn't have ports to talk to the printer. It also removes any chance of the printer running out of data due to a slow connection. Some printers with slots still need a computer with a host program to pick the right file from the card and start printing. Once the printing starts, you can disconnect the computer or turn it off.

Earlier, you read that you can put your start up and end Gcode into a program like Repetier Host and not have to put it into Slic3r. However, if your Gcode is going to go on an SD card that will print directly, you'll definitely need to put the start and end codes into your slicing program so that they will be on the card.

Model Considerations

Imagination is a wonderful thing. Part of the allure of a 3D printer is the ability to imagine something, draw it, and then see a realization of it form on the printer. However, imagination has to be tempered with some practical concerns. When you create a model for printing (or select one from some library of models) there are a few things you need to consider to ensure the model is practically printable.

Suppose you want to print a model airplane, for example. This poses several problems. If you think of the airplane in its normal orientation, it will almost certainly not be printable. The landing gear on a typical plane won't provide enough support for the entire body of the airplane.

Does this mean you can't print an airplane? Not at all. It just means you have to be smarter about how to print it. Generally, things with a larger footprint in contact with the print bed will produce the best results. It may be you can print the plane upside down, for example. Or it may be necessary to design the model in parts where the bottom is in contact with the print bed and the landing gear is separate and assembled after printing.

Overhang would be another problem. Unless the wings lay flat against the print bed, they will pose a similar problem. You could print them separately, of course. However, most slicing programs can generate support material to solve this problem (discussed above).

Depending on your printer and your print material, you should be able to print a certain amount of overhang if the angle is not too steep. Think of it this way: You can easily print, say, a cylinder. There is no overhang involved since each layer rests entirely on the previous layer. However, you can print an inverted cone by making each successive layer a bit larger than the previous one. The new layer will partially rest on the previous layer and, presumably, be hardened by the time the next layer is set down. The exact degree you can get away with this will depend on your printer, the material, and if you have active cooling (that is, a fan). Anything over 45 degrees, though, is highly unlikely.

There is one exception to the print over empty space rule. The printer can print bridges. A bridge occurs when the print head deposits plastic between two supported structures that have a gap between them. For example, suppose you have a part that has a groove in it. One way to print this would be to have the groove at the top so the printer would build the base and then the side parts of the groove. However, that could introduce other overhangs or printing problems. If the groove is reasonably narrow, the printer can lay down the side parts and then deposit a layer over the groove. The filament will droop a little, but over a few layers it will level out. The keys to printing good bridges is moving the head quickly and cooling the filament as quickly as possible. Lower temperatures help as does active cooling to make the best bridges.

The resolution of the model is also important. It is tempting to design (or download) models with fine lines such as lettering or details. However, there are definite limitations to any printer based on the nozzle size, the layer height, and the ability to cool the plastic quickly once it is out of the hot end. In particular, small or fine features may droop if the plastic doesn't cool fast enough. This is especially a problem when trying to print very small features that are the only thing on a layer. For example, imagine if you have a part that is a cube which is topped with a 1mm diameter pin that is several millimeters thick. The cube is no problem, of course. The pin will not require much time for each layer so the next layer may be trying to rest on plastic that isn't cool yet.

Slicing programs can arrange to turn on a fan or slow down printing for small features like this. The other option is to print multiple pieces. If the print head has to move to another part, this will give each part time to cool before the next layer prints.

A PC with no operating system isn't very useful. On the other hand, an operating system CD with no hardware isn't going to do much, either. Similarly, a 3D printer is a fusion of mechanics, electronics, software, plastic, and 3D models. Each part has a part to play, and a weakness in any component will show up in some way in the printed output.

6 WORKFLOW WITH AN EXISTING MODEL

If you haven't heard the term workflow before, it refers to the steps you take to get a desired outcome. For 3D printing, your workflow is pretty straightforward:

- Design or find a 3D model as an STL file
- Slicing the model into a GCODE file
- Initial preparation of the printer
- Send the GCODE file to the printer

Of course, exact details will vary depending on your exact set up, but this Chapter will cover the basics of the typical workflow. Chapter 5 covered the first two steps as well as the last one. This Chapter focuses on getting a typical printer ready to print.

Initial Setup

You don't always have to set the printer up, but you will, of course, the first time you print. You will also have to adjust the settings on occasion like when you load new filament, for example. Some printers require special setup for each print. For example, some printers require a one-time use print surface, and so you have to install the print surface before each print. However, most printers do not require per part set up.

Before printing, it is a good idea to inspect the printer. The bed should be clean of any plastic residue or oils. A scraping tool like a putty knife or a razor blade can get most of the plastic off a bed surface. Washing with acetone (for ABS plastic) is also a good idea. Alcohol can remove oil from your skin or oil or grease that accidentally contacts the print bed from the mechanics.

Depending on the material you are printing, you may need to check your bed material. For ABS you may have Kapton (a trade name for polyimide) tape that should be intact. People often print PLA plastic on blue painter's tape which you may have to replace from time to time. Nylon prints stick to burlap or other fiber surfaces. There are other surface treatments used with different plastics. For example, Aqua Net hair spray and Futura acrylic floor polish is known to help ABS to stick to a bed.

If you do use hair spray, be careful not to gum up the printer's mechanisms with it. You might consider spraying it into a container and then applying the resulting liquid using a rag or swab. Spraying on a hot bed causes quite a smell, so you might want to apply it while the bed is cold.

In general, nothing on the printer should move without the corresponding part of the printer's drive mechanism moving with it. If you can wiggle the hot end or make the X carriage move without moving the drive belt or screw, then something is not right. If you have unwanted motion, you should fix it before printing.

The hot end should not have burned residue or melted plastic sticking to the end. A fingernail emery board or a small file can be useful to clean a hot end. It can also help to heat up the hot end to soften the plastic before cleaning.

Filament Loading

Of course, before you can print, you need to load filament (or whatever print media you use, if your printer doesn't use filament). Some printers take a cartridge which makes it simple to load the plastic. Most low cost printers simply feed plastic filament directly. Filament usually comes on a spool of some sort or, if in small quantities, air spooled (that is, just wound in a coil). Your printer will have some mechanism to grab and feed the filament into the hot end. A Bowden extruder's feed isn't mounted on the moving part of the extruder. The filament is pushed through a tube that feeds into the hot end. An ordinary extruder mounts the feed part on the extruder head. The advantage to a Bowden feed is that it takes up less room and puts less weight on the moving part.

Your printer's feed mechanism may have some adjustment to control how tight the feed mechanism attaches to the plastic. Too loose and the filament may slip. Too tight and the gripping mechanism may chew into the plastic. Once you load plastic you may have to adjust the extruder feed calibration number in your software. This is done by using your control software to feed a known amount of filament (for example, 10mm) and using a caliper to measure how much it actually fed. An easy way to do this is to affix a small piece of tape to the filament and measure the distance from the tape to a fixed part of the printer. After feeding the filament, measure again and see how much the tape actually moved. For example, if the first measurement was 80mm and the second measurement was 65mm, then the filament actually moved 15mm. For this example, suppose the E calibration was at 820:

New_calibration = (820 * 10)/15 = 546.667

This makes sense if you think about the numbers involved. 820 is the number of steps you think is in a millimeter and you went 10 millimeters so that's 8200 steps. However you really went 15 millimeters so the steps per millimeter are 8200/15. You plug that back into your startup Gcode:

```
M92 E546.667
```

The best thing to do is to set the calibration to something first. Alternately, you can query the printer to see what is the default (most host programs can do this, or you can manually send Gcode to find the answer). However, you are going to change it anyway, so just set up the start Gcode, run a quick print that you can cancel right away, and then proceed with your calibration.

It is very important that the extrusion calibration is correct. If there is too much feed, your part will get too much plastic per unit volume which will fill in holes and throw dimensions off. Too little feed can cause voids in parts. This is also why you need to tell your slicing software the exact diameter of the filament (measured with a caliper).

You also, of course, need to calibrate the X, Y, and Z axis of your printer. However, in many cases the printer is already set to the right values. Follow your printer's instructions on how to set the calibration for the three axes. It is possible to adjust the exact number in the same was you did for the extruder. However, this isn't a good idea. A properly functioning mechanical system should be exact. For example, if your Z axis does 10mm per revolution on a threaded rod, and your stepper motors do 3200 steps per revolution, then your Z axis calibration is 320 (that is, there are 320 steps in one mm). If that doesn't work out, for some reason, you should examine the printer and figure out why. The extruder is a special case because the feed of the plastic isn't exact like the other parts of the printer.

Bed Leveling and Z Height

Another item that is difficult to get correct at first with most printers is the Z height. In a perfect world, the print bed would be completely level. The print head would also be completely level and move parallel to the bed. At the lowest point, the head should just float over the bed at somewhat less than the layer height above the bed. You'd like the plastic to squish a little bit on the bed so that it sticks. You don't want it to be so close to the bed that it can't get out of the print head, either. It makes sense that with a thinner layer height, the head will have to get closer to the bed without actually touching it.

In the real world, your bed is probably not level. You could, of course, use a level to be sure it is flat and straight. But a level will only tell you that the bed is level to the Earth. What you really care about is that it is in the same plane as the print head. So unless you have a plan to make both the bed and the print head level to the ground, you are going to have to do some adjusting.

You can use a feeler gauge to measure the height of the print head (at its lowest) over the bed. Using about half of the layer height is probably a good start (that is, for a 0.2 mm layer, you'd like the head to be about 0.1 mm above the bed). However, for most printing you can use a sheet of ordinary paper instead of a feeler gauge. You want to adjust the head and the bed so that you can pull a sheet of paper out from under the head and feel just a slight resistance. If you have to yank the paper hard, you are too low. If you don't feel a little scratchiness from the head as you pull the paper, you are too high.

The real issue, of course, is that just doing this at one point is not sufficient. Ideally, you'd like to start at the lowest point and adjust the head down until you have the right setting. Then you can move to other points and raise the bed. Sometimes you will have to iterate several times to get everything just right. Of course, your printer's bed may be fixed or the head may be fixed, at which point you have to adjust as best you can.

You might think this adjustment isn't very critical, but it is. If the plastic falls out of the head and doesn't squish on the bed, you will wind up with a tangle of filament that isn't sticking to the bed. If you jam the head too far on the bed, the plastic will not flow out and you will eventually jam the hot end or wear the filament down at the feed point. Either way, the plastic won't flow out and your printer will merrily keep tracing the lines of your model without leaving any plastic.

There is an art to finding just the perfect spot. Even a spot that works may have problems. If the plastic squishes a lot, it can leave a little skirt around the object and interfere with the dimensions. If the plastic doesn't squish enough, it may not stick to the bed and it may move, especially after a few more layers are on top of it.

Go Ahead: Print!

In Chapter 5, you learned how to download an STL file (or create one with a CAD package), slice it, and send it to the printer. With the printer checked out, you can load your file into your printer interface (or on an SD card) and start printing.

Prints can take a long time, so it is tempting to leave while the machine is working. However, you probably want to at least watch the first layer to make sure it works. Any problem with the first layer sticking to the bed will most likely ruin the print in the long run.

There are other things that can ruin your print. With experience, you'll be able to identify exactly what problem is causing your prints to go badly. That's the topic of Chapter 7: what causes prints to fail.

7 TROUBLESHOOTING PRINT PROBLEMS

When I was a kid, my parents ran a printing shop. Most of what we printed was business cards and it was done on an old letter press that would have not surprised Gutenberg. Today, we think nothing of printing with a high speed laser printer and getting perfect prints out in seconds. That wasn't the case with the old letter press. It was as much art as science. You had to get the platen aligned with the type and it was often a half hour of fiddling with it until you had an acceptable print.

Today's inexpensive 3D printers are more like a letter press than a laser printer. You can expect to spend some time tweaking and adjusting to get the best print quality. Only the more expensive printers even approach "plug and play." This Chapter will cover a few of the most common problems and their solutions.

Problems tend to come in several flavors: mechanics, unwanted motion, adhesion problems, and material issues. Usually, once you get the mechanics settled in, you shouldn't have many problems in that area. The same goes for unwanted motion (things like a bed that wobbles). Adhesion problems, however, can occur at any time and are often difficult to track down because there are so many possible causes for them.

This chapter assumes you have a correctly functioning printer to start with. However, if you've just built a printer, it is possible you'll find some help here as well. You do have to look a little deeper, though. For example, if your circles aren't coming out round and you know the printer should be working, it may be that your X or Y calibration is wrong (or, even, both). This could still be the problem with a new printer. But it is also possible you have the wrong kind of stepper motor, or the stepper motor controller is set to do a different number of steps than you expect, etc.

Sometimes it is useful to print the same part more than once even though it is failing. Something that fails at exactly the same spot each time indicates there is something wrong that is very specific. The file may be damaged, or you have a spot on a threaded rod that sticks. If the failure occurs at random, it is likely something that isn't so easily identified like a wrong temperature or a motor slipping.

You can also deduce a lot by looking at the axis that failed. If you see a rough layer alignment on two opposite sides of cube, you can assume there is something wrong with the perpendicular axis.

Of course, the cases in this chapter are somewhat generic. Your specific printer may have its own peculiarities. However, the most common problems are pretty universal, at least for a plastic filament printer.

Filament Doesn't Feed

If your filament doesn't seem to want to feed, there are several possible causes. The best thing to do is to first reload the filament and try extruding manually with the head several inches up over the bed. You should get a clean flow out of the hot end. The plastic will probably curl up, but that's not unexpected in this case. If you can extrude a fair amount like this, that pretty much means your problem isn't in the system from the hot end back to the filament.

In that case, the most likely cause is that the Z axis is starting too low. With the hot end jammed against your print surface, the plastic can't come out and eventually jams or gets worn enough that the part that moves the filament can't grip it anymore. Reset your Z home position and try again. Notice that once the filament gets worn or jammed, you will have to reload before you can try again.

If, however, you can't extrude into the air, you probably have a problem with the hot end, the extruder, or the filament. Things to check include having the right size filament, having a clogged hot end nozzle (rare, but possible), or having too much plastic dust in the extruder body to allow the mechanism to grab the filament. Most extruders have some sort of spring-mounted gripping mechanism. It is possible that it is not tight enough. Of course, the extruder motor must be turning along with the extruder gears. It is also possible that the extruder calibration is just so far off that the plastic is getting worn before it can be significantly moved through the hot end.

One check is to try manually pushing the plastic through a heated hot end. If that works, you can pretty much rule out a clogged head. The same goes for the extruder motor and gears. You can usually visually see if that's a problem or not, right away.

If the filament feeds at first, but eventually clogs, you might check the alignment between the feeding mechanism and the hole that enters the extruder body. On some printers, for example, there is a hobbed bolt and it should line up exactly with the hole the filament is supposed to go through. The tension idler should also line up. If these things don't line up, the filament may wear against the edge of the hole, causing poor results.

One other possibility for filament misfeeds is the extruder motor can't provide enough torque. Make sure the filament is free to feed (not caught on something, for example). The E stepper pot could require adjustment.

Filament Doesn't Stick

This is perhaps the most likely issue with printing. The print looks fine but won't stick to the bed. On the first layer or two this just causes a mess. If it happens later, it moves the whole print around and still causes a mess.

The most likely cause for this is your print head is starting too high. You need to align the Z axis to start at just a bit over the bed. Too low and the filament will jam (see above). Too high and it won't stick.

Another possibility is that your bed temperature is too low or the bed surface is either dirty or improperly prepared. If you use tape (blue tape or Kapton tape) make sure it is clean and flat. Some people rough the surface with a fine sandpaper first. For ABS Futura floor wax or Aqua Net hair spray are both effective at getting better bed adhesion. Some people use ABS juice when printing with ABS which is scrap ABS dissolved in acetone. A thin film on the bed will provide a good surface to print on, although you will have to trim up the base of the part once it is done.

Sometimes filament doesn't stick because the hot bed gets cooled down by something else. Many printers are enclosed in a box to prevent drafts. Also, printing in a cold room can cause the heater to not operate evenly or effectively. If you are printing with a fan, make sure the fan isn't cooling the bed too much. Most slicing software can disable the fan for the first few layers if you ask it to do so.

Filament Balls Up

This is really the same issue as filament not sticking, but it happens when the bed is way too high, or the bed is slick for some reason (that is, contaminated with oils, for example). Make sure your Z axis is adjusted and follow the same advice as filament not sticking, above.

Curling

Curling at the edges is usually due to the same reasons as filament not sticking. Some small parts can be very difficult to hold onto the bed. In that case, you might investigate if your slicing software offers a raft or a brim. A raft causes a thin layer to be printed before your part is printed. A brim causes a "lip" to be printed around the part. Either way, you gain extra surface area to hold the print down.

Corners Round Out

When your printer tries to make a corner, the plastic may drag and cut across the intended corner. This is a bed adhesion problem (see Filament Doesn't Stick above). You can see an example of this in Figure 7-1.

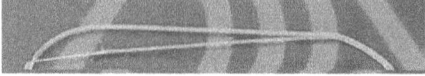

Figure 7-1. Rounded out corner

Edges are Rough

No printer is perfect, so some rough edges are to be expected. However, with a properly aligned printer it should be very minimal. If the edges are staggered only on one axis, suspect something is wiggling on that axis. It could be the hot end wiggling in that direction, or it could be backlash in the belt or rod that drives that axis. If the edges are rough all around, this could still be something causing the hot end to wiggle in all directions (see Figure 7-2). It could also mean you are over or under extruding. The volume of plastic flowing depends on the E stepper calibration, the nozzle diameter set in the slicing program, and the filament diameter set in the slicing program. These all need to be correct. Keep in mind that filament isn't always exactly its rated size.

Another thing to consider is speed. In general, slower speeds result in higher quality surfaces and edges. There is some point of diminishing returns and there is probably a ridiculously low point where the hot end burns the plastic because it is moving so slowly, but that would be very slow indeed. Of course, something that is wobbly may wobble less at a slower speed, so just because going slower makes it better doesn't necessarily mean there isn't a mechanical problem.

Figure 7-2. Rough edges from a wobbling hot end

Surfaces are Rough

Rough surfaces are typically due to trying to print too fast or an extrusion problem. The volume of plastic flowing depends on the E stepper calibration, the nozzle diameter set in the slicing program, and the filament diameter set in the slicing program. These all need to be correct. Keep in mind that filament isn't always exactly its rated size.

Print is Offset/Shifted

If a print seems to be good, then suddenly shifts on one axis and continues at least a bit more where everything would be correct if it had not shifted, that's an indication that one axis is slipping or stalling. This could be due to a loose belt or even a belt that is too tight (which can stall a stepper). If a bearing is binding, the offset will tend to happen in the same spot on any given layer. It is also possible that the trim pot on the stepper driver motor board needs to be adjusted.

Temperature is Unstable

If your bed or hot end temperature fluctuates, you probably have a draft or a breeze blowing against it. A faulty or loose thermistor (the temperature sensor) can also cause this. If the printer is new, make sure you don't have the bed thermistor swapped with the hot end themistor. A cooling fan can be a good thing, but you may need to turn it down or adjust its flow if you can't keep a stable temperature.

Some printer firmware will have a PID autotune feature (usually an M303 GCode, if your firmware supports it). A PID (Proportional Integral Derivative) is a type of control algorithm used to control the hot end temperature. The algorithm takes three coefficients (P, I, and D) that control its operation. Usually, the firmware defaults should be good enough for you, but if not, you may have tune them either automatically or manually, depending on your printer firmware.

Axis Only Moves in One Direction or will not Home

If you can only manually move one axis in one direction, this almost always means that a limit switch has a broken wire or has otherwise failed. Most machines have limit switches that are normally closed. So when a connection breaks—a very common failure mode—the printer assumes the head is always at that limit. This prevents slamming the limit, but it also means the printer thinks it is in the home position on that axis no matter where it really is.

Holes/Circles aren't Round

This is a common problem and typically indicates that at least one axis is out of calibration. For example, if the X axis is perfect and Y axis is off by 10%, a circle with radius of 10mm will come out as an ellipse with a 10mm radius in the X direction and an 11mm radius in the Y direction.

You can check your calibration and you could recompute the calibration so that it is correct. However, if the number is what your hardware should support, you should try not to change the calibration constant and instead figure out why the printer isn't matching the expected calibration (for example, a stretched belt).

Outside or Inside Dimensions are Off

If you print a calibration piece, it will usually have some expected dimension around the outside. If these are off, it is usually due to a bad calibration. You can check your calibration and you could recompute the calibration so that it is correct. However, if the number is what your hardware should support, you should try not to change the calibration constant and instead figure out why the printer isn't matching the expected calibration (for example, a stretched belt).

Another thing that can affect dimensions is over or under extruding. The volume of plastic flowing depends on the E stepper calibration, the nozzle diameter set in the slicing program, and the filament diameter set in the slicing program. These all need to be correct. Keep in mind that filament isn't always exactly its rated size.

If the outside dimensions are good, but the inside dimensions are off, it is most likely due to improper extrusion. It is always worthwhile to check both possible causes.

Layers Skipping, Squashing Together, or Hot End Drags

Sometimes a layer will start printing and it is obvious that the print head did not move completely off of the previous layer. This is most often caused by the Z axis slipping. However, it can also be caused by improper Z calibration or over extruding. In the latter case, too much plastic flows and the layer becomes too thick. Check the usual suspects for this: E stepper calibration, the nozzle diameter set in the slicing program, and the filament diameter (also set in the software).

Small Parts won't Stick

Even when you don't have a problem with large surfaces sticking to the bed, small pieces can be a real problem. Perhaps you are printing small separate items or perhaps it is a part with a detached small part (like the heel of a lady's shoe). The best way to avoid trouble is to make the part bigger, if possible. More surface area will stick better. However, that isn't always an option.

With ABS, you may be able to use ABS juice (ABS dissolved in acetone), Aqua Net hair spray, or Futura floor wax to increase the sticky properties of the bed. However, in many cases the answer is to print a brim or a raft (you may only have one of these options in your slicing software). A raft will cause a thin sheet of plastic to print under your entire print. If you are printing a single small piece, consider printing multiple copies at once so the raft will be large and the small parts will ride on the raft.

If you select a brim, it will print something similar to a raft, but only around the edge of the part. You can control the width of the brim, and you will want to make it wide enough to grip the bed. In either case, you'll need to trim the excess plastic after printing with a hobby knife, a file, or trimmers. Figure 7-3 shows two small parts with a brim around each.

Figure 7-3 Two parts, each with a brim

Bubbles or Voids in Surface

If the flat surface that was against your print bed appears to be bubbled, you probably don't have a perfectly smooth print surface. This is common with polyimide surfaces that get air bubbles trapped beneath them (similar to a phone screen protector). You really do want all the bubbles out. Figure 7-4 shows what happens with even small bubbles in the surface.

Figure 7-4. The back of a keychain printed on a surface with bubbles

If the bubble appears on the top of the part, it could be moist filament or a high extruder temperature. You can dry filament using desiccant (crystal cat litter works well) or with very low heat (but be careful not to get hot enough to deform the plastic or its spool).

If you have gaps that don't appear to be bubbles, low extrusion is the most likely candidate. Check your E stepper calibration and make sure extrusion isn't slipping or jamming.

Stringy Filaments on Print

If you find your prints have filament strings, you probably need to change the slicer's retraction setting. When the head is about to move without extruding, retraction causes the E motor to suck some plastic back up to prevent oozing. When extrusion resumes, the slicer will automatically extrude an equal amount to restart the flow. Figure 7-5 shows Slic3r's settings for this on the Printer Settings tab. Note, you don't need to set the "Extra length" parameter as this would cause even more plastic to extrude before restarting.

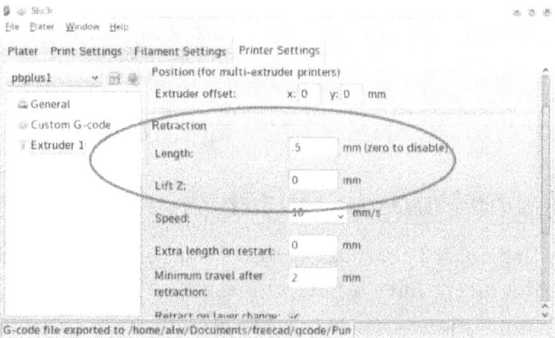

Figure 7-5. Slice3r settings that may help prevent ooze

The Lift Z setting can also help. This causes the head to lift up slightly before moving. Of course, it then moves down again before resuming. If your Z axis is adjusted right and you are not over extruding, this should not be necessary. However, if the head is a little low or you have too much plastic, the head may pick up a little plastic with it as it moves. Pulling it straight out will help minimize this and prevent the head from dragging.

If you do turn on Lift Z, keep in mind this will significantly increase Z motion. Be sure to keep an eye on your Z axis lubrication to avoid excessive wear on your printer's Z system.

One other case sometimes accounts for stringy prints. If you are trying to print "in the air" during a print, you will wind up leaving strings. This could happen because a bridge drooped too far, for example. Another common case is when a small piece of support (either deliberate support or a small piece of the actual part) doesn't stick. Subsequent layers in that area just leave a string of plastic hanging and when the head moves it will eventually deposit that string somewhere. The solution to this is to not print in the air.

There will always be a bridge too far to print, although you can try moving faster and cooling with a fan to increase your chances of a good longer bridge. If small support structures aren't sticking, see (Small Parts Won't Stick, above).

Obvious Ridge on Surface

If you can see an obvious ridge on a surface or on the sides of your print, this is caused by a small amount of plastic oozing out at the start or end of an extrusion. There are a few things you can do about this. Severe oozing may require retraction and Z axis lift (see Stringy Filaments on Print, above). However, some oozing is probably inevitable, but it shouldn't be enough to notice if it doesn't line up (see Figure 7-6).

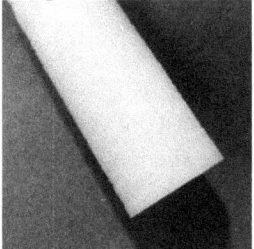

Figure 7-6. A foot with a vertical ridge caused by ooze

Most slicing software has a setting that allows you to randomize where extrusions start and end. For Slic3r, for example, the setting is on the Print Settings tab under Layers and perimeters (see Figure 7-7).

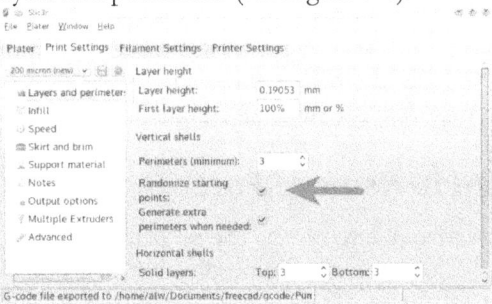

Figure 7-7. Slic3r settings to combat ridges

Print Overhang Collapses

When printing overhangs—that is an area of the part that isn't totally supported by the layer below—it is possible the overhanging part will collapse. This isn't really an error, *per se*. Printing at a cooler temperature and using fan cooling may help. Ultimately, though, some overhangs are going to require support material. You can build support into your design, or have the slicing software add it automatically if the angle is too steep.

Burning Plastic

If plastic is burning and leaving a black residue, that's a serious problem. If the plastic is burning inside the hot end, it may clog and that is an indication the temperature is too high. Cleaning a clogged hot end is usually a mess. You may have to disassemble it and soak it an solvent appropriate for the plastic you are using (acetone for ABS, for example).

If the burned parts are showing up outside the hot end, this could be because the head is dragging the surface (see Layers Skipping, Squashing Together, or Hot End Drags, above). Another possibility is that the extrusion temperature is too cool (or a fan is cooling the extruded plastic quickly) and it isn't clearing the hot end nozzle which then burns the plastic.

Plastic residue that sticks to the hot end nozzle will burn over time. An emery board (usually used as a fingernail file) is a good tool to quickly scrape a hot end clean of residue.

Plastic Makes Popping Noises/Lets Off Steam

This is almost always caused by damp filament. Keep filament sealed until you are ready to use it. If you can store filament with desiccant, that's even better. Storing it in a sealed container with silica gel (inexpensive if bought as cat litter) will keep it dry and even (eventually) dry out moist filament. If you decide to heat the filament to dry it, be careful not to use enough heat to soften the filament or melt the spool it is on.

Very Hard to Remove Part

If you have trouble removing the part from the bed, your Z axis may be down a little too far in the home position. This is a trade off because it means you probably don't have curling and bed adhesion problems. If you are using ABS juice to stick ABS to the bed, you can expect it to be very difficult to remove parts from the bed. Again, though, you are unlikely to have bed adhesion issues.

Figure 7-8 shows a part that has some squashing in the center. A hobby knife cleared the holes enough to allow the part to function properly.

Figure 7-8. This building block stuck to the bed near the center

When All Else Fails

Sometimes you have a problem and you simply can't find it. Here's a quick checklist:

- Try a different model to see if you maybe have a bad STL or GCode file
- Try reslicing the model, possibly with different settings to see if you've found a bug in the slicer
- If you haven't used the filament before, check to see if it is the size you think, that it is round, that it is dry, and see if has contamination on the surface or in the filament; try different filament
- Check your power supply; brown outs can cause mysterious problems that are hard to figure out
- Take a break or have someone else check over the printer with fresh eyes

8 POST PROCESSING

It would be nice if you could press the print button have a part magically appear and then be done. Sometimes it works like that, but more often than not, you'll need to do a little work to get there.

Post processing usually requires a few different tools. A hobby knife or knives is essential. You may find a rotary tool (like a Dremel tool) useful. A small set of files can also come in handy. Depending on your plastic, you may need some chemicals, as well. In general, you want a solvent, a glue that sticks to your plastic, and possibly some sort of pigment or paint.

Gluing

Gluing is one of the most common post production steps. Sometimes you will want to glue a part that broke off (perhaps it didn't adhere or maybe you got rough prying it off the bed). In some cases you will deliberately print a part in several pieces to get flat surfaces for printing and then glue the pieces together to form the whole.

Different glues work best with different plastics. For ABS, you can actually "weld" your part with acetone. The acetone will melt where it touches and the melted plastic will stick to other ABS. Of course, you have to be careful not let it melt areas you don't want it to glue. You can also use "ABS juice" which is ABS dissolved in acetone as a sort of paste glue.

Cyanoacrylate glue (commonly known as super glue or CA glue) will hold many plastics. Epoxy products like JB Weld can also bond with many plastics. There also special epoxies (like PlasticWeld) made to bond plastic. You can also actually weld plastic with heat to soften the material, although it can be difficult to keep the part from deforming in the heat.

Loctite publishes a document (Design Guide for Bonding Plastics – LT-2197) that has a lot of good technical information about types of glues and their bonding properties with plastics. At the time of this writing, you can download the document at:

http://www.henkelna.com/us/content_data/237471_LT2197_Plastic_Guide_v6_LR7911911.pdf

Of course, that URL could change. The henkelna.com site is the Loctite industrial site, so you should be able to find the current version if you start there.

Another adhesive you should not overlook is hot glue. A hot glue gun usually doesn't make a pretty bond, but if the bond isn't cosmetic (that is, it is hidden or on a part that you don't care what it looks like) it can make a fast strong bond. Be careful with low temperature plastics that the hot glue gun doesn't melt your work.

One final tool you'll want for gluing is clamps. For most work simple spring loaded clips will work better than C clamps, although they will get the job done as well. A bench vise, locking pliers, and even rubber bands will work in a pinch to hold parts together long enough to bond.

Sanding/Filing

Sanding or filing can clean up rough edges and smooth out surfaces. However, it doesn't take much to grind away most plastics. You want fine sand paper or emery paper. A rotary tool with a sanding head can also be very helpful.

For most parts, a big file will be hard to handle. You can easily get small files (needle files; see Figure 8-1) aimed at model makers and these are usually better. You'll find a wire brush will clean the plastic from the teeth, which will be important, especially with finer files. Needle files usually come in a set with different shapes. With practice, you can use round files to finish curves and even cut small notches with triangular files. A triangular file is also useful for making a square hole more square (a square hole coming out roundish can be a sign of over extrusion). You should have plenty of scrap plastic to practice with if you save waste from your prints that don't work out. In a pinch, a nail file or an emery board can fill in.

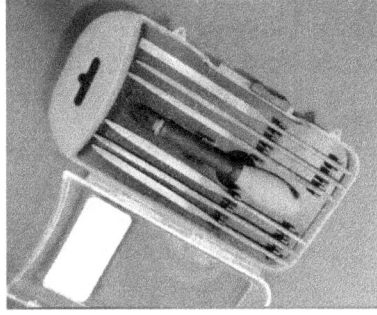

Figure 8-1. A set of needle files

Another consideration while filing or sanding is that you are generating plastic dust. Some of this could be harmful or, at least, irritating. Work where it is ventilated and, if possible, wear a dust mask.

If you have access to a tumbler or a sandblaster, either of these can be carefully used to surface finish a part. Shaking in a bag of abrasive substance might offer slight improvement to your surface finish.

Fill

Related to sanding, it is possible to use modeling putty to fill in voids just as you might work on an automobile body. Once the putty dries, you can sand or otherwise remove the excess to get a smooth finish. In my experience, I'm not artistic enough to make this work, but some people get great results with this method, especially if the finished product is painted to cover the difference in materials.

Mechanical Finishing

A very common operation is to drill out a hole that doesn't quite fit. A small electric drill with an assortment of bits is handy for this. You may also find a rotary tool will get the job done.

Side cutters or a hobby knife can be used to trim excess plastic or remove support material. Keep in mind, though, that cutting into colored plastic may leave a blemish on the color finish.

If you decide to use mechanical fasteners to hold your pieces together, you have at least three choices:

- Design in a clearance hole to take a nut and a bolt
- Design a nut trap that will hold a nut internally and use it to engage a bolt
- Use wood screws to tap into the plastic

Instead of wood screws, you may have better luck with sheet metal screws, depending on your plastic choice. There are also special plastic-threading screws known as "thread-forming screws." These work well, but aren't strictly necessary. Suppliers include Screwerk (http://screwerk.com/), TR Fasteners (http://www.trfastenings.com) and doubtlessly many others.

Painting

Regular paint doesn't usually stick to plastic very well. However, you can find paints from brands like Rust-Oleum and Valspar that are specifically for plastic. The paint has a solvent in it that actually bonds the pigment to the plastic.

Another similar technique is to use nail polish. This is also usually solvent-based and packaged for fine detail work. Acrylic paint has also been reported to stick to ABS and PLA. For small detail work, a permanent marker (such as a Sharpie) will color most plastics.

For some plastics, like PLA, you may be able to use a primer like gesso before painting with ordinary paint. Depending on your plastic and your paint, you'll have to experiment to see what works best for you.

Instead of painting, dyes can add color to projects. You can use commercial dyes, or you can use tea or red wine to give an aged color. For example, see http://www.shapeways.com/model/146656/universal-ring-dial-2.html For an example that uses tea and ink to give a very distinctive appearance.

Besides painting, you may want to try clear coating parts. This adds to durability and can also improve appearance.

Heating

A hot air gun (see Figure 8-2), a hair dryer, or even a small torch used carefully can improve surface finish. PLA, in particular, can be treated with a flame (quickly, as PLA melts easily) to improve the surface finish and color.

Figure 8-2. A typical hot air gun

Hot plastic can also be deformed to fix a warped piece or introduce a bend into a piece that was too hard to print in the bent shape. Just remember to be careful not to overdo it and be sure to have sufficient ventilation and just in case equipment should you set a fire with a torch.

Chemical Treatment

There are many chemical treatments you can use to attempt to make your prints "better" in some way. For example, ABS plastic can be dipped in acetone. This will soften the surface and melt the obvious layers away. It also causes loss of detail as the whole surface softens, not just the layers. Most people report better results if they dip the part in acetone and then let it air dry.

Another common approach is to use heat to create acetone vapor. This is a fairly risky proposition since acetone is flammable and you should breathe the fumes, so I won't discuss it more here. The idea, though, is that by depositing a fine and even layer of acetone, the vapor is like a dip with less deformation. Presumably, any plastic could use a solvent dip or vapor bath with the appropriate solvent. ABS happens to work well with acetone.

Instead of acetone, some brands of carburetor cleaner have become popular for treating both ABS and PLA. Reportedly, it makes ABS surfaces smoother and will make PLA flexible, even after drying. I haven't tried this myself, but you can see a video about it at http://youtu.be/lP-S3PkvBbk.

About Post Processing

Don't get too hung up on the post processing of parts. Sure, you always need to trim off a little excess or enlarge a hole. That's just part of printing. But the complex acetone baths, paints, and heating are really for people who are trying to make things beautiful. If you are building artwork or gaming figures or model railroad scenery, you may want to go all out and try to do as much finishing as you possibly can.

However, if you are building repair parts or commonplace objects, you may be surprised at just how good a well adjusted printer can do without much effort after the part comes out of the printer. If you have multiple heads, adding multiple colors can really make designs look appealing. If you don't have multiple heads, consider printing multiple pieces and joining them to make attractive everyday objects.

Figure 8-3. A binary clock (files at
http://www.thingiverse.com/thing:113334)

For example, consider Figure 8-3. This is a binary clock (well—an analog binary clock). There are three printed parts: the ring, the face, and the base. You can print these in different combinations of contrasting colors. The ring holds the face with a little hot glue on the rear of the clock.

I also had to enlarge the clock hole in the center because the movement I bought at a local store was a little bigger than the original design called for. Painting the numbers on the face or the entire face and the numbers would make it even more attractive.

Al Williams

9 THE FUTURE

As a ham radio operator, I always envied the old guys when I was a kid. They had a certain excitement about radio because they had been there in the very early days of it. To them, our modern gear (well, modern when I was a kid) was nothing short of a miracle. I have no idea what they'd think of what we have today.

These days, I'm an old guy when it comes to computers. My first computer was painstakingly built and had a keypad (not a keyboard) and some LED displays. The computer I'm writing this on would have made NASA green with envy in those days. So I feel lucky to have that miraculous feeling when I work on today's computers.

I also think I've gotten lucky twice. We are in an era where 3D printing is starting to take off. When I built my first computer, it was very much like building a printer today. Big companies had had them for awhile, but they were hard to use and very expensive. If you wanted one, you better be prepared to learn all about it because you'd have to do everything yourself from the building of the power supply to the programming of the I/O routines. The results weren't spectacular, but we loved it and I can honestly say I got as much—or maybe more—satisfaction from programming those little hand-made machines as I do today working on what amounts to a supercomputer on my desk.

Imagine if 3D printing parallels this path. The computers I'm talking about are only about forty years old. What will 3D printing be like in forty years? It boggles the mind.

Already we are seeing experiments in printing with liquid metals and conductive polymers. There is news almost daily about printing microscopic parts and printing in Earth orbit. New printers have multiple heads, support plastic, and—if you can afford it—self-leveling beds. Even today's most expensive printer will seem as archaic as a TRS-80 or other old computer in a few years.

What fun! We get to be the trailblazers. We get to develop new techniques and better ways of doing things. The field is wide open and growing at a dizzying rate. In a few decades some mega-corporation will own the space and do all the important (and expensive) research. But today, it can be us.

I'm glad to be here.

I hope this book has given you some insight into what a printer is, how it works, and what you can do with it. I tried to balance and give you details you needed without assuming you were using a particular brand or type of printer.

If you decide to start printing with a 3D printing service, you'll want to find out what materials and capabilities they have. If you buy your own printer, you'll need to read some specific documents to find out about what it does and how to operate it. My goal wasn't to show you how to use my printer. I wanted to get you thinking about what you would do with your printer.

Hurry up. We have a future to invent.

APPENDIX I: USEFUL LINKS

Updated lists at http://www.hotsolder.com/3dprint

Information and Models

http://reprap.org/wiki/Main_Page - The original web site that started it all.

http://www.thingiverse.com – Many 3D printable models

http://grabcad.com/home - Sophisticated 3D models (many not printable)

http://sketchup.google.com/3dwarehouse/ - Sketch up models (many not printable)

Vendors

http://www.makerbot.com/ - Possibly the best known 3D printer maker in this market

http://printrbot.com/ - My personal choice of printer

http://www.ultimaker.com/ - Award winning printers

http://trinitylabs.com/ - Maker of the Aluminatus printer

http://www.solidoodle.com/ - Inexpensive assembled printer

http://www.lulzbot.com – Popular printer vendor claims low layer heights and high speeds

http://makibox.com/ - Inexpensive printers made from acrylic

Print Services

http://www.shapeways.com – Prints in a variety of material

http://i.materialise.com/ - Another printing service

http://www.sculpteo.com/en/ - Print from an uploaded model

http://www.ponoko.com/ - Prints uploaded models

http://www.kraftwurx.com/ - Cloud-based 3D printing

CAD Software

http://www.tinkercad.com - Excellent browser-based 3D modeling

http://www.netfabb.com – Fixes and optimizes models for 3D printing

http://www.3dtin.com – Another browser-based modeling tool

http://shapesmith.net/ - Another browser-based modeling tool

http://www.openscad.org/ - Parametric modeling software

http://brlcad.org/ - Modeling software originally built by the U.S. Army

http://www.sketchup.com/ - Google 3D software

http://sourceforge.net/projects/free-cad/ - Open source professional-style 3D CAD

http://www.123dapp.com/catch - Photos to 3D models

http://www.blender.org/ - Artistic modeling

APPENDIX II: GETTING STARTED WITH OPENSCAD

A complete treatment of OpenSCAD is beyond the scope of this book. However, there are so many times that describing a simple part parametrically is useful, that you should at least have some familiarity with the tool. Also, Thingiverse supports a simple way to create "customizers" that allow you to provide an OpenSCAD script to other users that they can customize without any knowledge of OpenSCAD on their part.

Getting Started

OpenSCAD runs on your Windows, Mac, or Linux PC. You can find installation instructions on the OpenSCAD web site (http://www.OpenSCAD.org/downloads.html). When you run the program, it doesn't look like a normal CAD program (see Figure A2-1).

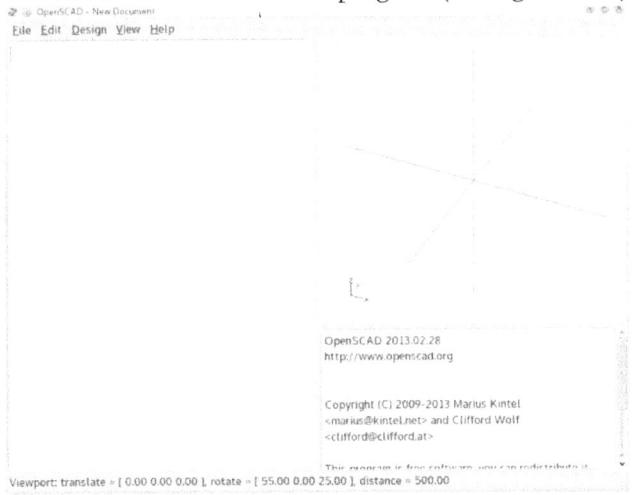

Figure A2-1. OpenSCAD main window

You will enter what amounts to a program or script on the left hand pane of the window. Then you'll use the Design menu to render that program into the top right pane. You can code a little and then preview your drawing. If something isn't right, you can fix it or you can wait until you've completed everything to try to fix things. When you are completely done, you can render the output in different ways, including as an STL file.

A Simple Panel

Suppose you want to make a simple front panel for an electronic gadget. The panel is 125mm wide by 50mm high and should be 2mm thick. There are four 5mm holes for LEDs centered on the panel. There are also 4 #6 bolt holes for mounting 8 mm from each corner.

It is easy enough to make a blank panel by entering the following language into the top left pane:

```
cube([125,50,2]);
```

Enter that and press F5 (or use the Compile item from the Design menu). The result should look like Figure A2-2. You can use your mouse to change the view. Left dragging rotates the view. Right dragging pans. You can use the scroll wheel to zoom in and out.

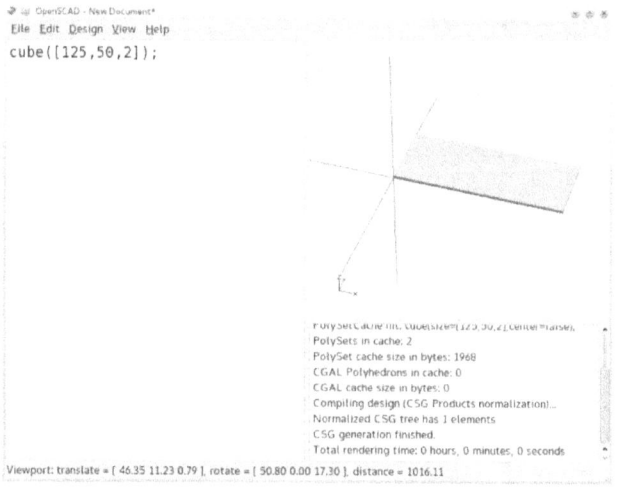

Figure A2-2. A simple panel

Just a blank panel isn't very exciting though. Also, since the numbers are hard coded, you'll have trouble later. Let's fix that first:

```
panelWidth=125;

panelHeight=50;

panelThick=2;
```

```
cube([panelWidth,panelHeight,panelThick]);
```

This changes nothing, but it makes it much more readable. Later, you'll use these variables in other places and it will make things simpler to change. For example, let's add the LED holes. First, add the following line to the bottom of the ones that are already there:

```
cylinder(h=panelThick*2,r=5,center=true);
```

When you press F5 again, you'll be disappointed. The cylinder is right at the corner of the panel. Not to mention, you need a hole and this is another solid. We can fix the first problem by translating the cylinder to a new position. Change the cylinder line so that it reads:

```
translate([panelWidth/5,panelHeight/2,panelThick/2]) {

cylinder(h=panelThick*2,r=5, center=true);

}
```

When you press F5 now, you'll see the cylinder has moved to where the first LED hole should be. You may have to tilt the view a bit with the mouse to see it. Because it is tied to the variables you set up earlier, if you change the size of the panel, the hole will move to the right spot. The problem is, it isn't a hole, it is a cylinder poking out of both sides of the panel (thanks to moving the Z axis to panelThick/2). That's easy to fix with a difference operation. Here's the entire file (so far):

```
panelWidth=125;

panelHeight=50;

panelThick=2;


difference() {

cube([panelWidth,panelHeight,panelThick]);

  translate([panelWidth/5,panelHeight/2,panelThick/2]) {

  cylinder(h=panelThick*2,r=5, center=true);

  }

}
```

The difference operation takes the first thing (the cube) and makes it solid. Then it subtracts all the other things until it sees the matching curly brace (note the translate command also has curly braces). In this case, there is only one other thing, and the result is the panel with the LED hole cut out of it.

Of course, this is just one hole and you need 4. That's easy to add with a for loop:

```
panelWidth=125;

panelHeight=50;

panelThick=2;


difference() {

cube([panelWidth,panelHeight,panelThick]);

  for (i=[1:4]) {

    translate([panelWidth/5*i,panelHeight/2,panelThick/2])

    {

    cylinder(h=panelThick*2,r=5, center=true);

    }

  }

}
```

The bolt holes at the edge are similar. A #6 bolt hole has a clearance hole of 0.1495 inches. Some more cylinder commands will do the trick:

```
panelWidth=125;

panelHeight=50;

panelThick=2;


difference() {

cube([panelWidth,panelHeight,panelThick]);

  for (i=[1:4]) {

    translate([panelWidth/5*i,panelHeight/2,panelThick/2])

    {

    cylinder(h=panelThick*2,r=5, center=true);

    }

  }

// bolt holes #6
```

```
translate([8,8,panelThick/2])

    cylinder(h=panelThick*2,r=0.1495*25.4,center=true);


translate([8,panelHeight-8,panelThick/2])

    cylinder(h=panelThick*2,r=0.1495*25.4,center=true);


translate([panelWidth-8,8,panelThick/2])

    cylinder(h=panelThick*2,r=0.1495*25.4,center=true);


translate([panelWidth-8,panelHeight-8,panelThick/2])

    cylinder(h=panelThick*2,r=0.1495*25.4,center=true);

}
```

Note that if there is only one thing after the translate command, you don't need the curly braces. The // characters start a comment and causes the system to ignore the rest of the line. Figure A2-3 shows the finished design.

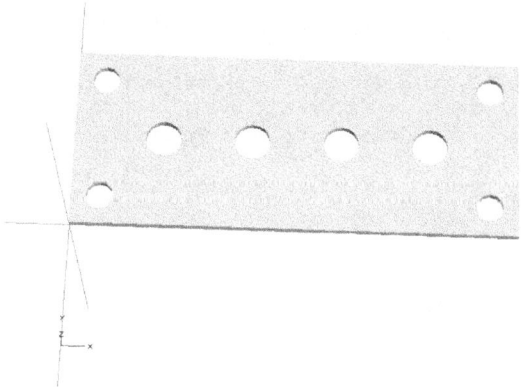

Figure A2-3. The finished panel

When rendering, OpenSCAD has to determine how many line segments to use when making circles (like the ones in our holes). There are three special variables that control how OpenSCAD decides how many fragments to use when making a circle. The first is the $fn variable. By default, this is zero, but if you set it to something other than zero, then OpenSCAD will use that as the number of line segments to make a complete circle.

When it is zero, then the $fa (default 12) and $fs (default 2) variables come into play. The circle has, of course, 360 degrees and a particular circumference. OpenSCAD will use the smallest of 360/$fa or $fa times the circumference. With the default values, that means a circle will have 30 segments unless the circumference divided by 2 ($fs) is smaller. However, OpenSCAD always uses at least 5 segments.

A circle with just a few segments won't look very good (a circle with 4 segments is a square; try setting $fn=4 at the start of your file). You should set $fn to a high value early in the file. You can also override these values in a particular call such as:

```
cylinder(h=5, r=2, $fn=18);
```

There is a lot more to learn about OpenSCAD. The documentation is a good place to start. If you are impatient, you might enjoy the OpenSCAD cheat sheet at http://www.openscad.org/cheatsheet/.

Thingiverse Customizer

When you share an OpenSCAD design on Thingiverse, you can have the web site customize your OpenSCAD file based on user input. The process is simple. Consider a plate with some #6 bolt holes in it. I use these to clamp belts, but you might use it anywhere you need a "mending plate." However, you might need different sizes or numbers of holes.

The trick is to use variables with special comments. Here's the mending plate file:

```
/* Really simple "mending plate" with two #6 holes

    -- I use this to clamp a GT2 belt,

       but it probably could be used

    for anything. Al Williams, April 2013 */

// Length of plate (mm)

length=17; // [10:100]

// Width of plate (mm)

width=6; // [5:100]

// Thickness of plate (mm)

thickness=3; // [.5:20]

// Offset from edge/hole spacing (mm)

edgeoffset=3.5  ;  // [1:10]
```

```
// Number of holes (should be even)

n=2; // [[1:100]

loopct=n/2;

difference() {

cube([length,width,thickness]);

for (i=[1:loopct]) {

  translate([edgeoffset*i,width/2,1])

      cylinder(h=thickness*10,r=2,center=true,$fn=100);

  translate([length-edgeoffset*i,width/2,1])

      cylinder(h=thickness*10,r=2,center=true,$fn=100);

}

}
```

The web site will offer input for any variables that get a constant value. If you want "hidden" variables, you can add 0 to the constant to prevent the system from picking them up. In the example, loopct is the only variable that is hidden and it isn't set up from a constant anyway.

Figure A2-4 shows an example of customizing the plate on the Thingiverse web site.

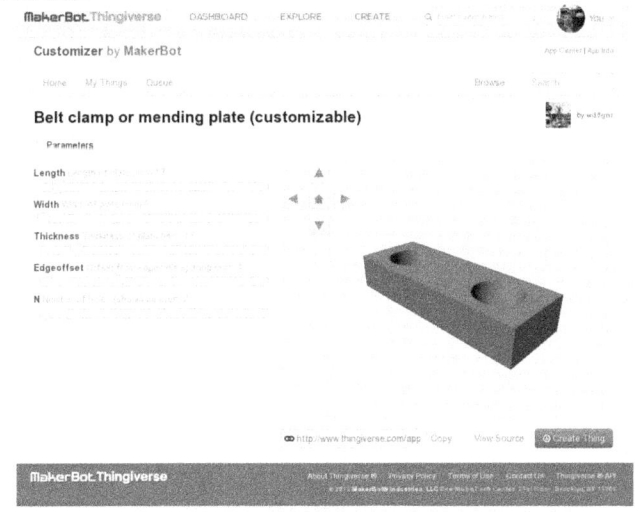

Figure A2-4. Customizing the mending plate

The comments directly before the variable set the explanatory text. The numbers in square brackets set the range of choices. If the range has a colon, the range turns into a slider (like the ones in Figure A2-4). If it is a comma-separated list, the customizer will create a drop down box. If you don't put any range at all, the customizer will use a regular input box. There are several other syntax tricks, but these will take care of most of your designs. You can read all about your choices at: http://www.makerbot.com/blog/2013/01/23/openscad-design-tips-how-to-make-a-customizable-thing/.

APPENDIX III: FORMULAS

Background

Most calculations need to know a few basic items:

- Steps per revolution – most steppers will do 200 steps per full revolution although this is not always true
- Controller microstepping – Most controllers will break a single step into 16 or 32 microsteps
- Pitch – Belts and leadscrews have a pitch. For a belt this is the number of teeth per millimeter. For a leadscrew it is how many millimeters the threads move for a full turn
- Pitch on leadscrews vary by type. For example, an M6 screw has 1 millimeter per turn. A 5/16" -18 screw has about 1.4111 millimeter per turn. Note that if your drive system uses gears, you will need to take the gear ratio into account with the formulae below.

Inches to millimeters

(inches) = (millimeters) / 25.4

Millimeters to inches

(millimeters) = (inches) x 25.4

Microsteps per Revolution

(microsteps per revolution) = (motor steps per revolution) x (microsteps)

Degrees per Microstep

(degrees per microstep) = 360/(microsteps per revolution)

Millimeters per Microstep (belt)

(millimeters per microstep) = ((pulley tooth count) x (belt pitch)) / (microsteps per revolution)

Millimeters per Microstep (leadscrew)

(millimeters per microstep) = (leadscrew pitch) / (microsteps per revolution)

Microsteps per Millimeter (either)

(microsteps per millimeter) = 1 / (millimeters per microstep)

Stepper Motor Calibration

(new calibration constant) = (old calibration constant) x (commanded distance) / (measured distance)

BONUS: COPYING A FLAT SOLID OBJECT

In this bonus chapter, I'll show you how I took a wooden letter "W" that had broken, took a photograph of it, digitized it, repaired it, and then make a 3D model out of it. You can find the files on Thingiverse at http://www.thingiverse.com/thing:43387.

The first job was to get a picture. It didn't need to be a great picture, but it did help to get it on a high contrast background. I just used a simple digital camera to take the picture (see Figure 1). Notice the break near the bottom left (just as the stroke of the letter starts to go towards the central peak).

Figure 1. The original wooden item (complete with break)

The problem, of course, is that the photograph is made of pixels, and the 3D printer wants a 3D model made up of triangles and vectors. The plan of attack goes like this:

- Use an image editor (GIMP) to convert the picture to a simple black and white image with clean edges (and the damage repaired, in this case).
- Use a vector image editor (Inkscape) to trace around the letter and produce an SVG file (a 2D vector format).
- Import that vector image into a CAD program (OpenSCAD) as

a geometry. Use the CAD program to "extrude" the 2D path into a 3D object.

- Print the STL file and presto, a new perfect W is born. The results appear in Figure 2 (a small test print and a full size print).

Figure 2. Two printed copies of the original (at different scales)

Obviously, this chapter can't teach you everything there is to know about GIMP, Inkscape, and OpenSCAD. But I will show you the basic steps I took and the commands I used to get the job done.

Images from digital cameras are made of pixels (a raster image). Inkscape uses vectors—instead of dots, the image is described as lines and directions. This is more similar to the way the 3D STL file format works. That's why the conversion has to go through Inkscape and not directly into OpenSCAD.

GIMP

Before you can convert your image into a vector format, it needs to be as simple as possible. The shape should be all one color (usually black) and the background should be a highly contrasting color (like white). Like most photo editors, GIMP provides a variety of smart selection tools. I used the fuzzy selection tool (the one that looks like a magic wand) to select the object. If I missed some spots, I held down the shift key and added more to the selection. In the case where I got too much, I used the control key to subtract a new selection from the existing one.

Keep in mind that you can start with the magic wand tool and then use other selection tools to add or subtract. For example, adding a rectangular selection was useful for repairing the break in the letter. If there is any shadow or any part of the vertical faces of the original in the picture, don't select them. You want the selection to be a representation of the flat face of the part. This also means when shooting the picture you should try to get the camera lens as close to perpendicular to the part as possible. By the same token, taking the picture against a solid-colored contrasting background will be very helpful as well.

Once you have the selection outlined, you can press the delete key to replace the entire selection with a solid background color. By the color selector in the toolbox there are some swap arrows that let you swap the foreground and the background color. You can press then and then right click inside the selection and pick Select | Invert. This will cause everything that is not the part to become selected. Hit delete again to change this background to a solid color (the original foreground color).

GIMP's native file format isn't useful in this process, but you might want to save a copy anyway in case you want to touch up the selection process. The working file, however, needs to be a JPEG and to do that you have to use the File | Export menu item, not the File | Save command.

Once you have the JPEG you've done most of the hard work. Figure 3 shows the end result of the processing with GIMP.

Figure 3. The photo converted to pure black and white

Inkscape

When you launch Inkscape it will start with a blank page. The page size isn't going to make any difference in this case, so don't worry about it. Start by importing the JPEG file you saved from the first step. The goal is to get a path around the outline of the part. This ought to be hard, but Inkscape has a Path | Trace Bitmap command (see Figure 4) that will do the work automatically. Because of the work done in GIMP, the settings on the trace command aren't especially critical. You can play with them until you get a good looking path. You can use the Update button to get a preview or you can simply press OK and then undo anything that wasn't satisfactory.

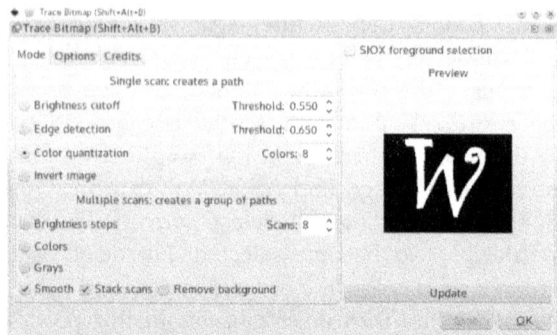

Figure 4. Inkscape's trace bitmap dialog

Once you press OK and you are happy with the result, you can press the Edit Paths button (or press F2) and select your new path. You'll see lots of nodes lit up. You can delete any extraneous nodes (you can even select lots of them with a drag and delete them *en masse* if you like). You can also grab the nodes and pull or push them to modify the shape of the path. When you are satisfied you might want to try Path | Simplify to reduce the number of nodes.

You may find it easier to delete the JPEG under the path after tracing, or you may like to leave it there while you work for reference. However, before you save the file as a DXF (a standard CAD file format) you need to delete the bitmap and leave only the path.

In theory, you can save the file using Inkscape's "Desktop Cutting Plotter" format (you can find all the formats on the Save As dialog). However, the export in this format is a bit too simple, and most people prefer to use a third party add-on to save to DXF. One that works well is provided free from Big Blue Saw (http://www.bigbluesaw.com/saw/big-blue-saw-blog/general-updates/big-blue-saws-dxf-export-for-inkscape.html). You might also try the one at http://www.thingiverse.com/thing:25036 although recent versions of Linux have trouble running this one. However you do it, you need to save your work as a DXF. Again, you might also save a copy as an SVG (Inkscape's native format) in case you want to go back and touch it up again.

If you wanted a cookie-cutter effect (that is, just the outline of the letter or shape) you could copy and paste the path so that there were two copies. Change the fill color of one so you can tell them apart. Then drag the copy so that it is exactly on top of the original. Use the Path | Inset command to "shrink" the copy a few times until you have the thickness you want (you'll be able to see because of the different colors—the original color will be the edge.

Once it is how you want, select the original and then use the shift key and select the copy. Path | Difference will then subtract the copy from the original, leaving an outline suitable for export to make a cookie-cutter.

OpenSCAD

OpenSCAD can import DXF files. Older versions used an argument to the linear_extrude command to specify the file. Newer versions complain that you ought to be using the new way, but still allows the older method. Just in case, here's both ways to do it.

With older versions of OpenSCAD, this was the line of code required to read the /tmp/w.dxf file:

```
linear_extrude(file="/tmp/w.dxf", height=5,center=true);
```

However, the "new" way to do it looks like this:

```
linear_extrude(height=5,center=true)
import(file="/tmp/w.dxf");
```

The result should be the same. The file is converted to an OpenSCAD geometry and then it is extruded 5 millimeters to be a 3D object. If the object shows up as an outline, you didn't get down to two colors in GIMP (that actually could work if you wanted a cookie cutter). For this project, the result should be a solid object as seen in Figure 5.

Figure 5. The result in OpenSCAD

Once you have the line entered (setting your desired extrusion height and file name, of course), you simply press F5 or F6 (Compile or Compile and Render). Then you can execute the Design | Export as STL menu item to save your STL file.

Of course, you could just use the imported shape just like any other OpenSCAD object and build up something more complex. For example, consider this script which produces the output seen in Figure 6.

```
$fn=64;

union() {
```

```
cylinder(height=5,r=15);

translate([0.7,-8,0])

  linear_extrude(height=5,center=false)

    import(file="/tmp/w.dxf");

}
```

Figure 6. Adding shapes in OpenSCAD

Print!

Armed with the STL file, you can print just like any other model. You might need to adjust the scaling a bit to get the size to exactly match the original (if that's important). The ability to copy real objects with a digital camera is pretty powerful. Of course, 3D scanners exist and there is even software that can attempt to take multiple pictures and change them to a 3D object. Those are outside the scope of this book, but still an exciting possibility.

It is worth noting that you don't have to start with a picture of a real object. The same technique works fine if you want to draw something in Inkscape and then print it. That's how the object in Figure 7 was created. The text started out as an Inkscape object. You must convert everything to a path and then the rest of the process was just the same as printing the giant letter.

Figure 7. Inkscape text converted to plastic

INDEX

Al Williams

ABOUT THE AUTHOR

Al Williams is a long time electrical engineer, programmer, author, ham radio operator, and all around techie. He lives in Houston, Texas with his wife Pat, way too many computers, and just enough dogs.

www.ingramcontent.com/pod-product-compliance
Lightning Source LLC
Chambersburg PA
CBHW071238170526
45165CB00003B/1144